TECHNICIAN'S GUIDE TO INSTRUMENTATION

Technician's Guide to Instrumentation

by

Samuel Simons

Uhai Publishing, Inc.
1731 Helderberg Trail
Berne, NY   12023
Toll free 1-877-842-4782
www.uhaipublishing.com

NOTICE TO THE READER

Cover Design:
   Bi-County Graphics

Cover Photographs:
   courtesy AMETEK Calibration Instruments
   courtesy Fluke Instruments
   Photo Disc

Interior Photographs:
   courtesy Dora Kay Simons
   courtesy Fluke Instruments

Illustrations by:
   S. Simons/ Bi-County Graphics

ISBN: 1-930528-14-0

COPYRIGHT © 2004
UHAI Publishing, Inc.

Printed in the United States of America

For more information, contact:
Uhai Publishing, Inc.
1731 Helderberg Trail
Berne, NY 12023
Toll free 1-877-842-4782
www.uhaipublishing.com

1 2 3 4 5 6 7 8 9 10   XXX   08 07 06 05 04

Preface

This book is written to assist students in an Instrument Technology Degree Program or Certified Apprentice Training Program in acquiring the skills, knowledge, and abilities necessary to perform the typical assigned duties of a Journeyman Instrument Control technician (ANSI II). The material is presented in a practical and progressive application format. Troubleshooting skills and safety awareness are reinforced throughout the book. The physical properties of, and common measuring techniques for, pressure, temperature, level and flow are explained. The purpose, function, interaction, and application of primary measuring devices, transmitters/transducers, controllers, and final control elements are explained. Calibration methodology and technique including measuring/test equipment, test set-ups, documentation, error analysis, pass/fail operational assessment, and adjustments are detailed for each device.

Individual chapters are dedicated to safe shutdown devices and control valves. The text also includes detailed analysis of proportional band, integral, and derivative and explains their individual and combined tuning effects on single capacitance control loop. Several application scenarios are developed to reinforce the student's comprehension of the interplay between individual devices, control loops, and the process. A troubleshooting chapter guides the student through a controlled application of troubleshooting techniques. The final chapter reinforces skills and behaviors necessary for a safe and productive work environment. An instructor's guide is available that includes chapter overviews, review questions/answers, and test questions/answers.

__Chapter One__ introduces and explains the fundamentals of process control including components, signal transmission path, and control schemes. It defines and explains feedback, feed-forward, and cascade control, elementary/one line drawing, and piping and instrument diagrams.

__Chapter Two__ describes the physical principles and precautions pertaining to pressure, temperature, level, and flow, and explains the operation of primary element devices. It includes calculations to convert units and determine span, range, elevation, and density corrections. It also identifies operability testing techniques to determine primary sensing element malfunctions.

__Chapter Three__ details the calibration process. It covers the documentation process including standards, traceability, reports, and records; methods to obtain, analyze, and interpret test data; and adjustment techniques to correct errors found in pneumatic and electronic devices.

Chapter Four identifies and explains the selection, use, and operation of common testing devices used to simulate, detect, and identify pneumatic and electronic signal characteristics. It addresses precautions and hazards of connecting test equipment including foreign material, incompatible products, static sensitivity, and grounds.

Chapter Five describes the function, operation, and maintenance of pneumatic and electronic transmitters, relays, solenoids, LVDTs, and proximeters, including adjustment techniques and operability assessments. It also covers simple series, parallel, and bridge circuits.

Chapter Six explains the function, operation, and maintenance of chemistry analyzers. It also covers operability assessment and troubleshooting.

Chapter Seven describes the basic function, operation, and maintenance of pneumatic, electronic, and digital controllers. It explains control modes, transfers, PID response, tuning, calibration, operability assessment, and troubleshooting techniques.

Chapter Eight explains the function, operation, and maintenance of final control devices and support hardware. It covers application, analysis, tuning, calibration, operability assessment, and troubleshooting. It also lists several hazards and states the safety precautions associated with AOV calibrations.

Chapter Nine explains the application, function, operation, and intent of emergency shutdown systems (ESDs). It covers the process and machine variables typically monitored, and the methods used to effect interlocks, trips, and permissives. It also covers operability assessment and troubleshooting techniques.

Chapter Ten explains the application and function of programmable logic controllers (PLCs). It covers the hardware, inputs/outputs, binary numbers, ladder logic symbols, addressing, and logic operands used by a PLC. It also covers operability assessment and troubleshooting techniques.

Chapter Eleven explains the application and operation of a distributive control system (DCS). It covers basic hardware, protocol, tag data, and system topology as they relate to distributive control. It also covers common errors and troubleshooting techniques.

Chapter Twelve details the application and operation of field-mounted devices including, tuning, control transfer, and removing and restoring equipment to service. It also covers hazards and precautions, operability assessment, and troubleshooting.

Chapter Thirteen explains the systematic and ordered approach to troubleshooting. It differentiates and categorizes faults and failures, and explains symptom recognition, symptom analysis, system manipulation, and fault validation. It also covers root cause analysis, change analysis, and half-split bracketing.

Chapter Fourteen *explains hazards and precautions associated with work activities. It covers MSDS, DOT placards, confined space, personal safety equipment, and lockout/tagout. It emphasizes clear and concise communication techniques including teambriefs, the phonetic alphabet, and three-part communication.*

The Author, *Samuel L. Simons, is a third generation instrument and control technician. Over the last thirty years, he has served many facets of the instrumentation and control industry including apprentice, journeyman, master technician, trainer, supervisor, engineer, manager, and owner/operator. He has an Honors Degree in Electronics and holds the INPO National Academy of Nuclear Training certification both as an instrument and control technician and as a training specialist. Mr. Simons is a former delegate to the National Conference of Standards Laboratories and a member of ISA.*

Acknowledgements:

The author would like to thank the following people for their expert assistance and guidance in developing this text:

The WE team at South Texas Project

Megan Graham, Casper College

Bryan Harvey, South Georgia Technical College

Gordon Hobbs, Austin Industrial Company

Alan Houtz, Kenai Peninsula College

W. Richard Polanin, Illinois Central

Table of Contents

Chapter 1

Fundamentals of Control

OBJECTIVES

Upon completion of Chapter One the student will be able to:

- *List the fundamental components of a control loop.*
- *Describe the typical signal transmission path of a control loop.*
- *Convert equivalent linear scale values from pneumatic to current.*
- *Convert equivalent linear scale values from current to voltage.*
- *Differentiate linear, square root, and log scales.*
- *State the advantages of square root and log scales.*
- *List the common control schemes.*
- *List the common engineering prefix for notation values.*
- *List the inputs to an elementary or line drawing.*
- *List the outputs from an elementary or line drawing.*
- *Define process, control, instrumentation, set-point, algorithm, measured variable, manipulated variable, controlled variable, feedback, binary logic, truth tables, elementary/one line drawing, and piping and instrument diagram.*
- *Describe the purpose of process control instrumentation, a primary element, a transmitter, a controller, and a final control element.*
- *Differentiate between open-loop and closed-loop control, on-off and analog devices, series and parallel circuits, feedback, feed forward, and cascade control.*
- *Identify, interpret, and analyze the following: piping and instrument diagrams and symbols for field and control board instrumentation and interconnecting signal lines, Relay numbering code, elementary*

drawings and symbols for normally open/closed contacts and relay coils, And gate symbol and truth table, Or gate symbol and truth table, inverter gate symbol and truth table, and latch gate symbol and truth table.
● *Convert engineering units and attach the appropriate prefix.*

1.0 Introduction

The term *process* refers to a physical or chemical change of matter or conversion of energy. Control implies regulation or manipulation of the variables influencing a process in order to achieve a desired result. Instrumentation is a specific device or groups of related devices that measure, monitor, and/or control a process. The fundamental purpose of process control instrumentation is to measure, monitor, and control a process. For thousands of years processes were controlled completely by human manipulation. Skills such as firing clay pottery, baking bread, and smelting ore were passed on and handed down through the generations. A person or group of people directly observed the process, determined the control actions required, and manually manipulated the required variables to obtain a satisfactory product. Overall, efficiency was poor and product quality was completely subjective. These early manual processes were often trial and error, and the results, based entirely on the personal skill of the artisan, were inconsistent. Modern processes are, for the most part, completely automated, and require minimal human intervention. Efficiency is high and product quality is rigidly controlled. Results are predictable and repeatable. The predominant human role in process production today is to monitor and maintain automated control systems.

1.1 Fundamental Components of a Control Loop

The four fundamental components of a control loop are the primary element, transmitter, controller, and final control element. Figure 1.1-1 demonstrates the function and association of the basic elements of a control loop.

Primary elements detect and convert energy from the process variable into a quantitative value suitable for measurement. The measured variable is the process parameter monitored by the primary element. Transmitters receive and interpret the measurement value from the primary element, convert it into a standardized signal of representative value, and transmit the standardized signal to the controller.

There are several scales relating transmitter/transducer signals to board readings. The most common are linear, square root, and logarithmic. Scales convert actual process values into proportional analog transmission signals suitable for transmission, and back into indications of the actual process value for indication and recording purposes, see Figure 1.1-2. The primary element originally converts process energy into a detectable value for the transmitter. The transmitter, controller, and associated loop instrumentation process and

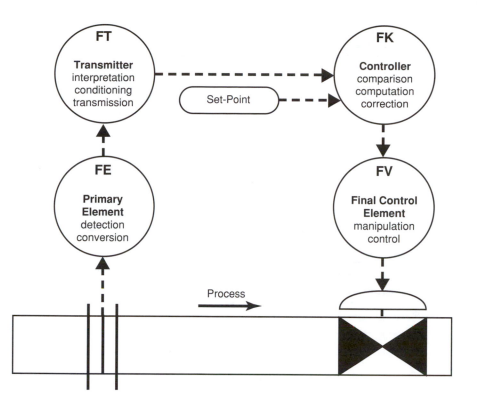

Figure 1.1-1 *Block Diagram of Simple Control Loop*

transmit the data as a standardized signal, such as 3 − 15 psig, 1 − 5 volts DC, or 4 − 20 mADC. The data is translated back into the original representative process value at the final indicating or recording point. The most common scale is the linear or percent scale. Linear scales are graduated to indicate an equivalent percentage of process value in relation to percent of instrument signal. Square root scales are graduated to extract non-linear flow values from a linear instrument signal. Logarithmic scales are commonly used in analytic processes. The advantage of log scales is the ability to maintain excellent resolution throughout an extreme range. Figure 1.1-3 compares several types of scales. Figure 1.1-4 details the input and output spans of a simple pressure loop. Figure 1.1-5 details the actual process value signal at 50 psig as it is conditioned and transmitted through the loop. The transmitter converts the 50 psig process value into an equivalent current value of 12 mADC. The controller receives 12 mADC from the transmitter and compares it to the set-point (50 psig). With no offset (error) the controller outputs 12 mADC to the I/P. The I/P converts the 12 mADC current signal into the pneumatic equivalent, 9 psig, and transmits it forward to the control valve. The control valve's available stroke is 0 − 2 inches from full closed to full open. Nine psig positions the valve at one-inch open. To convert linear scales calculate the span by subtracting the lower range value from the upper range value. A temperature transmitter with a range of 20 to 220°F has a span of (220 - 20) = 200°F. Divide the span by 100 to determine the number of units span equal to one percent (200 ÷ 100) = 2. In this instance two degrees of temperature is equivalent to one percent of the span. To determine the temperature value at 50% span, multiply the units per percent by the desired percent

Process Value	Percent Span	psig	mADC	Volts DC
100 psig	100%	15	20	5
75 psig	75%	12	16	4
50 psig	50%	9	12	3
25 psig	25%	6	8	2
0 psig	0%	3	4	1

Process Value	Percent Span	psig	mADC	Volts DC
1000 psig	100%	15	20	5
750 psig	75%	12	16	4
500 psig	50%	9	12	3
250 psig	25%	6	8	2
0 psig	0%	3	4	1

Process Value	Percent Span	psig	mADC	Volts DC
220°F	100%	15	20	5
170°F	75%	12	16	4
120°F	50%	9	12	3
70°F	25%	6	8	2
20°F	0%	3	4	1

Figure 1.1-2 *Linear Equivalent Scale Values*

(2 x 50 = 100) and add the product to the lower range value (100 + 20 = 120). Mid range of 20 – 200°F is 120°F. This value is converted into the pneumatic equivalent by obtaining the 50% value of the 3 – 15 psig pneumatic span in the same fashion. Subtract the lower range value from the upper range value to determine the span (15 - 3 = 12). Divide the span by 100 to obtain the units of value per percent of span (12 ÷ 100 = 0.12). Multiply the units per percent by the percent desired (0.12 x 50 = 6) and add the product to the lower range value (6 + 3 = 9). Fifty percent of 3-15 psig is 9 psig. Therefore 9 psig is proportional to 120°F. The same process is used to transpose all linear proportional scales.

It is important to note that all signal ranges have a lower range value that is greater than zero. This provides for a "live" zero condition. A process value of zero percent will always generate some signal value, such as 3 psig, 4 mADC, or 1 volt. A signal value below the signal's lower range value indicates a fault condition.

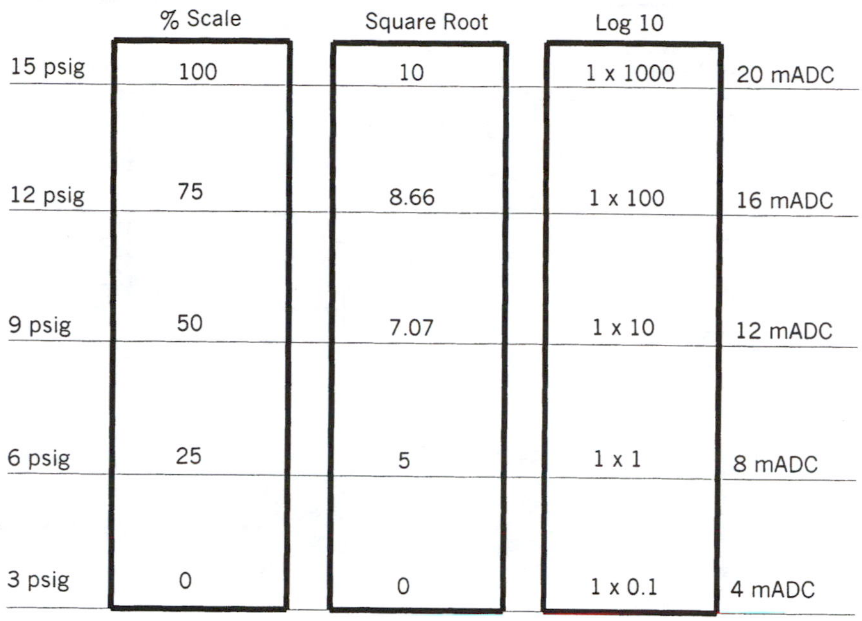

% Scale
Square Root
Log 10
15 psig
100
10
1 x 1000
20 mADC
12 psig
75
8.66
1 x 100
16 mADC
9 psig
50
7.07
1 x 10
12 mADC
6 psig
25
5
1 x 1
8 mADC
3 psig
0
0
1 x 0.1
4 mADC

Figure 1.1-3 *Percent Square-Root and Log Scales*

4 – 20 mADC
PK
4 – 20 mADC
4 – 20 mADC
I/P
3 –15 psig
Set-Point
0 – 100 psig
4 – 20 mADC
PI
0 – 100 %
4 – 20 mADC
PT
0 – 100 psig
3 – 15 psig
PV
0 – 2" Travel
Process
0 – 100 psig

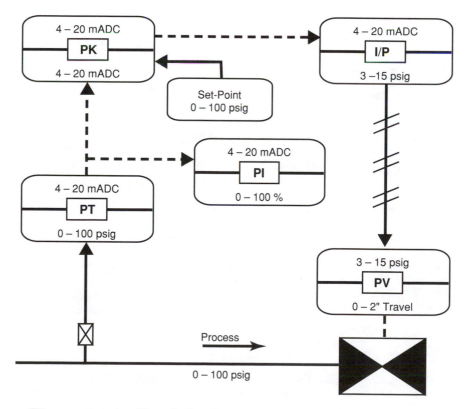

Figure 1.1-4 *Signal Conditioning and Transmission Path*

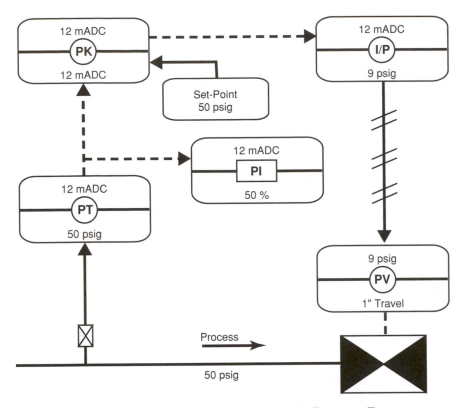

Figure 1.1-5 *Signal Values at 50 psig Process Pressure*

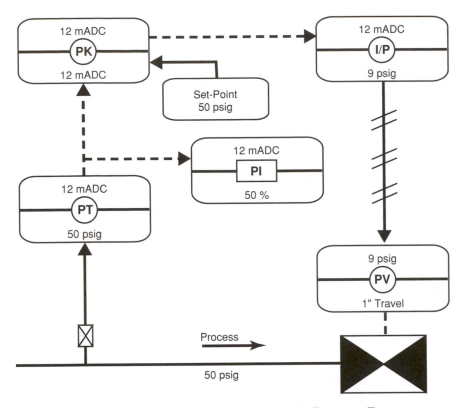

Figure 1.1-6 *Annunciator Panel*

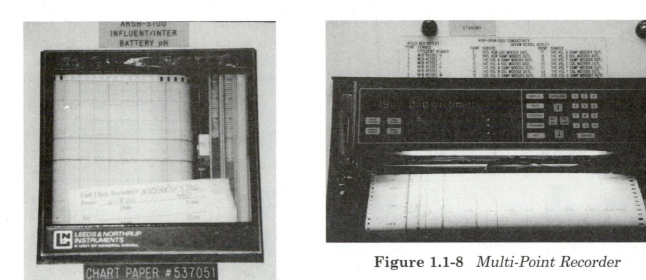

Figure 1.1-8 *Multi-Point Recorder*

Figure 1.1-7 *Two-Pen Recorder*

At this point, the signal is conditioned sufficiently for indication, recording, or alarms. The controller receives the transmitter signal. Controllers, operating automatically, compare the signal to the desired value (set-point) and apply preset algorithms to create a correction signal. The set-point is the input variable that the controller recognizes as the desired control value. An algorithm is a pre-established set of specific logical or mathematical functions or instructions applied to any given variable. The controller then transmits the correction signal to the final control element. Final control elements receive the correction signal (demand) from the controller and respond accordingly. Final control elements function to directly change the value of the manipulated variable. If the controller algorithm is correct, the measured variable returns to set-point and the loop is stabilized.

The manipulated variable is the process parameter directly altered by the final control element. Typical manipulated variables include flow, temperature, pressure, and level. The controlled variable is the process, parameter, or product over which a desired or influenced result is anticipated. In a simple feedback control flow loop the measured, manipulated, and controlled variables are sometimes the same variable. In most feed forward control applications the measured and manipulated variables are not directly associated with the controlled variable; influence over the controlled variable is incidental.

1.2 Open-Loop and Closed-Loop Control

A closed loop control system is any control system that has a feedback signal. Feedback is the ability to monitor the results of a control action and determine if additional correction is required. An open-loop control system is a control system operating without feedback. In Figure 1.2-1, the open-loop system attempts to maintain the liquid level in the tank by cycling the valve on and off at preset time intervals. If all variables remain constant, this is

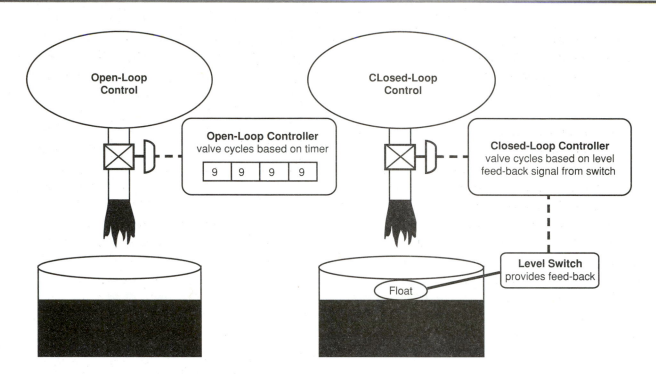

Figure 1.2-1 *Open-Loop/Closed-Loop Comparison*

an effective control technique. The closed loop system directly monitors the level in the tank and manipulates the valve as required to keep the level constant. Closed loop systems have the ability to automatically correct for disturbances. Figure 1.2-2 shows a simple closed loop feedback pressure control application. The valve is preset to remain at

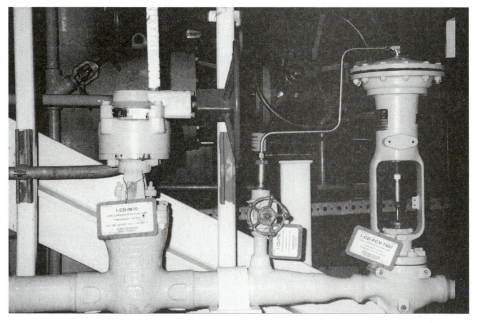

Figure 1.2-2 *Simple Pressure Control Feed-Back*

mid-position when the downstream pressure is at the desired value. If the pressure drops, the valve opens to allow more flow and increase the downstream pressure. If the pressure exceeds the desired value, the valve closes to restrict flow and decrease downstream pressure.

Many basic concepts are similar in all applications of control. Control schemes include a measured variable, a set-point, and a manipulated or controlled variable. A familiar control scheme, a temperature controller (thermostat) found in common central heating units, is depicted in Figure 1.2-3. The user selects a desired temperature (set-point). The controller compares the measured variable (room temperature) to the set-point. The controlling unit detects any existing deviation and transmits a correction signal to the final controlling element (heater coil). The heating unit incorporates on-off control. On-off control operates under one of two discrete states, ON or OFF. ON State is also referred to as (1) or TRUE, and the OFF State is referred to as (0) or FALSE. If room temperature is below set-point, the heater turns on. The heater stays on until the room temperature matches set-point, and then it turns off. A low oil pressure warning light is also an example of an on-off device. If oil pressure is below set-point the light is on; if above set-point the light is off.

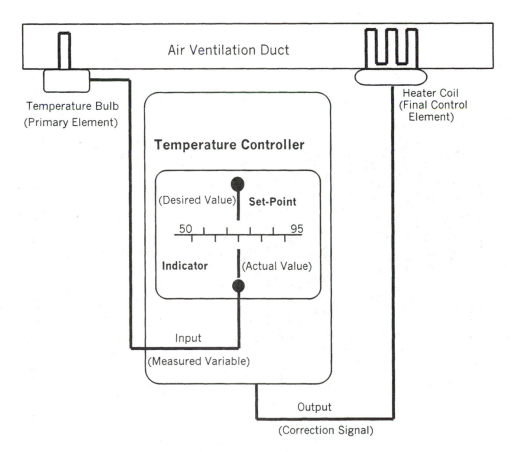

Figure 1.2-3 *Temperature Control Loop*

Figure 1.2-4 *Temperature Controller (Inside)*

Another familiar example of closed loop control is the cruise control system in a modern automobile. Again, the user selects a desired speed and the controlling system attempts to maintain that speed. Cruise controls use analog control. Analog devices are capable of receiving, manipulating, and transmitting an infinitely variable set of values. They are limited only by the upper and lower limits of their design and their degree of resolution or sensitivity. If automobile speed drops below the set-point, the controller will incrementally attempt to accelerate the automobile back to match the pre-set set-point speed. If the speed error is small, the controller outputs a signal that demands the automobile to accelerate gently. If the speed error is large, the controller outputs a signal that

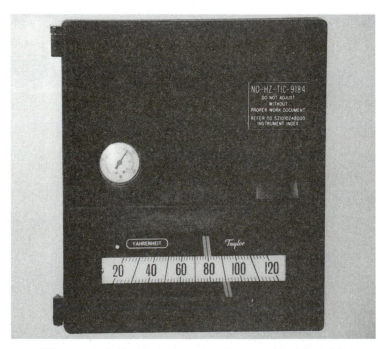

Figure 1.2-5 *Temperature Controller (Outside)*

demands the automobile to accelerate harder. An oil pressure gauge is another example of an analog device.

Application and design parameters determine the type of control selected. Analog control, although not generally used in residential heating, is acceptable for use in home heating units. However, if personal comfort is the main performance criteria, the additional cost and complexity add little to overall value. On-off control is a simple and inexpensive alternative for analog automobile speed control units. The ride, however, requires nerves of steel and considerable good fortune. The only control options available are accelerator full open and accelerator full closed.

1.3 Binary Logic

Binary logic is an application of mathematics that interprets, represents, and determines all possible values, states, or conditions based on a successive progression of TRUE/FALSE statements. Many process control systems are microprocessor based and operate in a digital format. Digital components process all information as a series of high and low states. A system of binary mathematics, Boolean algebra, is used to translate TRUE/FALSE logic into a string of ones (1) and zeros (0). TRUE is depicted as "one" and FALSE is depicted as "zero." A single digit of binary is referred to as a bit. Groups of eight bits are called a byte. The lowest order bit in a binary number (2^0) is the "least significant bit". The highest order bit in a binary number (2^n) is the "most significant bit."

Complex control schemes often contain a multitude of circuitry and components. It is often unnecessary or impractical to determine with certainty how each of the different compo-

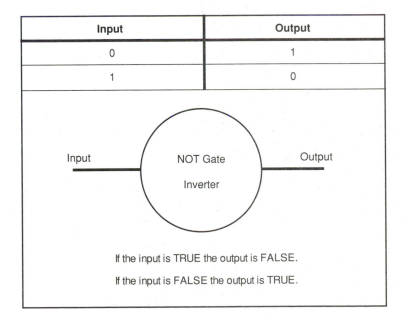

Input	Output
0	1
1	0

Input — NOT Gate Inverter — Output

If the input is TRUE the output is FALSE.
If the input is FALSE the output is TRUE.

Figure 1.3-1 *Inverter Truth Table*

Input A	Input B	Output OR	Output NOR
0	0	0	1
0	1	0	1
1	0	0	1
1	1	1	0

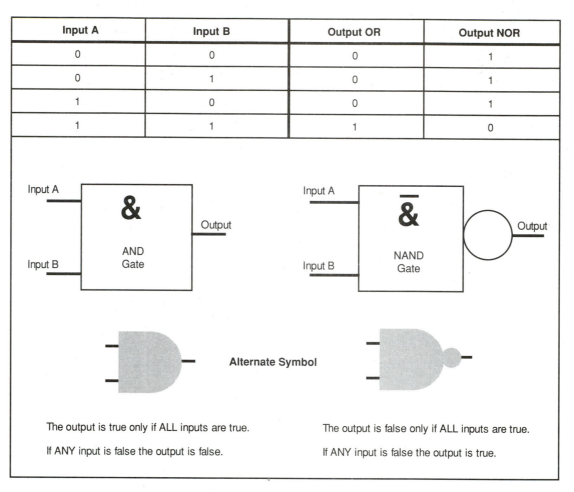

Figure 1.3-2 *AND/NAND Gate Truth Table*

nents interrelates. Many times the control scheme is represented by a set of logic diagrams that contain representative blocks, or gates, that map the functional operation of the system. There are only a few symbols used to represent the function of the system and the relationship between signals. The most common symbols are the AND/NAND gate, OR/NOR gate, NOT gate, and LATCH. The function of these gates is an application of Boolean algebra. Each gate performs based on a particular truth table for that gate. Truth tables are charts of each possible output for all possible inputs to a logic gate.

For any given input, an INVERTER (NOT GATE) outputs the logical opposite as in Figure 1.3-1. The AND gate only outputs a TRUE (1) if all inputs are TRUE, Figures 1.3-2 and 1.3-3. A FALSE (0) applied to any input of an AND gate will result in an output of FALSE. A NAND gate is simply an AND gate with an INVERTER on the output.

The OR gate outputs a TRUE (1) if any of the inputs are TRUE, Figures 1.3-4 and 1.3-5. The OR gate outputs a FALSE (0) only if all inputs are FALSE. The NOR gate is simply an OR gate with an INVERTER on the output.

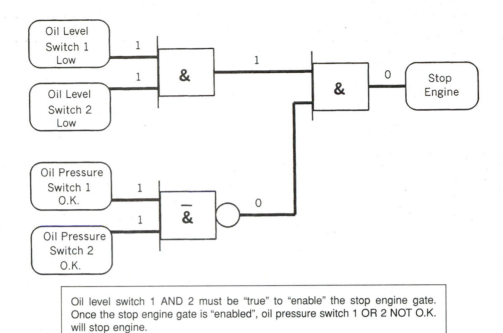

Oil level switch 1 AND 2 must be "true" to "enable" the stop engine gate. Once the stop engine gate is "enabled", oil pressure switch 1 OR 2 NOT O.K. will stop engine.

Figure 1.3-3 *AND/NAND Circuit*

Input A	Input B	Output OR	Output NOR
0	0	0	1
0	1	1	0
1	0	1	0
1	1	1	0

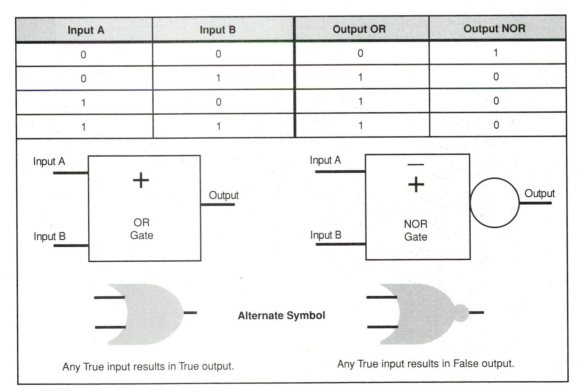

Figure 1.3-4 *OR/NOR Gate Truth Table*

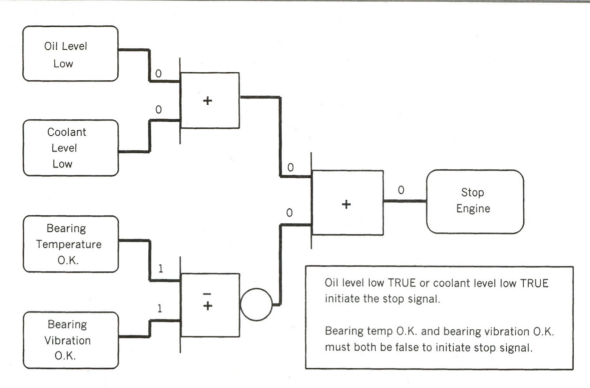

Figure 1.3-5 *OR/NOR Gate Circuit*

LATCH gates are used to catch and hold a logic condition, Figure 1.3-6. If both inputs are FALSE (0) the latch output remains CONSTANT at the last value. If the LATCH is RESET by logic TRUE on the RESET INPUT, the latch output, A, goes FALSE. The output will remain a logic (0) until the SET input goes TRUE (1). If both SET and RESET are

Input S	Input R	Output Q	Output $\overline{Q}$
0	0	No Change	No Change
0	1	0	1
1	0	1	0
1	1	Toggle	Toggle

Figure 1.3-6 *Latch (Flip/Flop) Truth Table*

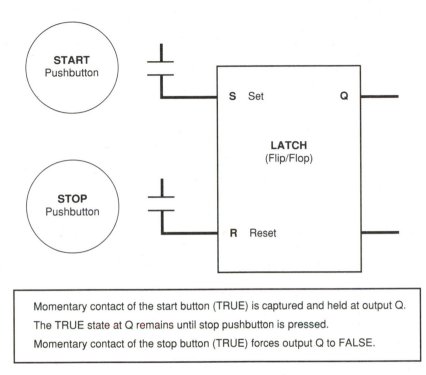

Momentary contact of the start button (TRUE) is captured and held at output Q.

The TRUE state at Q remains until stop pushbutton is pressed.

Momentary contact of the stop button (TRUE) forces output Q to FALSE.

Figure 1.3-7 *Simple Latch Circuit*

TRUE, the LATCH will TOGGLE. The TRUE output turns FALSE and the FALSE output turns TRUE. A common application for a LATCH is to monitor start/stop pushbuttons as in Figure 1.3-7.

1.4 Interconnecting Wiring Diagrams

Interconnecting wiring diagrams are drawings that depict control schemes as a system of series and parallel lines that represent the individual measurement and control elements of a process. Elementary and one line drawings also illustrate electrical wiring schemes. These types of drawings portray relay, "hard-wired," logic and control applications. Commonly used symbols, shown in Figure 1.4-1, are contact normally open, contact normally closed, coil, and timer. Electrical drawings often use "Electrical Power System Device Function Numbers" a code established by ANSI and IEEE. The complete code is located in the controlled reference publications ANSI/IEEE C37.1-1979 and ANSI/IEEE C37.2-1979 (or current revision). Figure 1.4-2 lists several common devices and their respective code numbers.

Elementary drawings begin the signal path at the load center or fuse box that supplies power to the particular loop. The power source is depicted as two parallel horizontal lines (runs) crossing the upper and lower sections of the drawing. The line (run) across the upper portion of the drawing is the input/power line. The lower line is the return line (run). Each

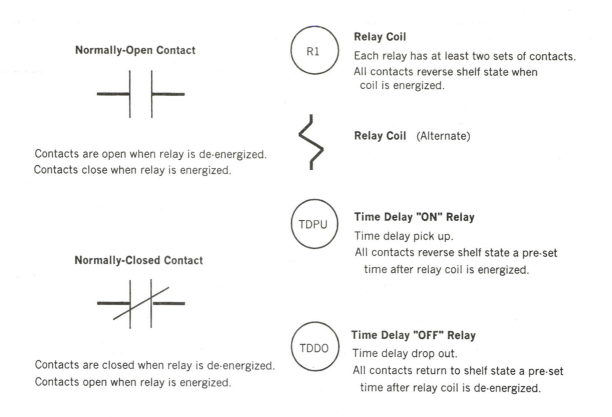

Figure 1.4-1 *Basic Relay/Elementary Logic Symbols*

individual circuit is represented as a vertical line (branch, leg, or drop) connecting the upper and lower lines. Components along the same vertical line are wired in series. Components on vertically adjacent lines are wired in parallel.

Series circuits provide only a single path or route to conduct signal transmission or current flow as demonstrated in Figure 1.4-3. If the path is interrupted at any point, all circuits on the path are de-energized. Signal transmission is only permitted if all contacts along the series path are conducting. Parallel circuits provide multiple paths for signal transmission as demonstrated in Figure 1.4-4. If one path is interrupted, the signal may continue (conduct) on an alternate or adjacent path.

Each branch in the circuit contains an input (switching device) and an output (coil/relay). The input is often a field contact from a pressure, temperature, level, flow, or valve position switch. Pushbutton contacts, hand switches, relay contacts, and timers are also inputs. Outputs drive alarms, energize solenoids, start/stop motors and pumps, and provide logic states back to other circuits. The operational sequence of the circuit determines the selection or placement of input and output devices on the branches as in Figure 1.4-5.

Device Number	Function	Description
1	Master Element	Sequence or operatiom initiate switch
2	Time Delay Start/Close	Provides desired amount of time delay
12	OverSpeed Device	Speed switch actuates on overspeed
14	Under Speed Switch	Speed switch actuates on underspeed
20	Electrically Operated Valve	MOV, SOV
30	Annunciator Relay	Visual/Audible alarm device
38	Bearing Protective Device	Bearing temperature or vibration
48	Incomplete Sequence	Logic discordance interrupt
62	Time Delay Stop/Open	Provides desired amount of time delay
63	Pressure Switch	Operates on pressure value or change rate
65	Governor	Control system that regulates operation of machinery or equipment
69	Permissive Control Device	Permits or inhibits equipment operation
71	Level Switch	Operates on level value or change rate
74	Alarm Relay	Provides visual/audible indication
80	Flow Switch	Operates on flow value or change rate
83	Transfer Relay	Automatically selects condition or equipment
86	Lock Out Relay	Abnormal condition interrupt and hold out
90	Regulating Device	Regulates voltage, current, power, speed, frequency, temperature, or load within close limits

Figure 1.4-2 *Electrical Power System Device Function Numbers*
(Refer to ANSI/IEE C37.1 and C37.2 current revision)

A and B = True

All Normally-Open contacts in series must change state (close) to conduct the signal.

$\overline{A}$ or $\overline{B}$ = False

Any Normally-Closed contact in a series path must change state (open) to interrupt the signal.

N.O. (Normally-Open)   N.C. (Normally-Closed)

Figure 1.4-3 *Basic Series Circuits*

A or B = TRUE

Either normally-open contact closing conducts signal.

$\overline{A}$ or $\overline{B}$ = FALSE

All normally-closed contacts must open to interrupt signal.

Figure 1.4-4 *Basic Parallel Circuits*

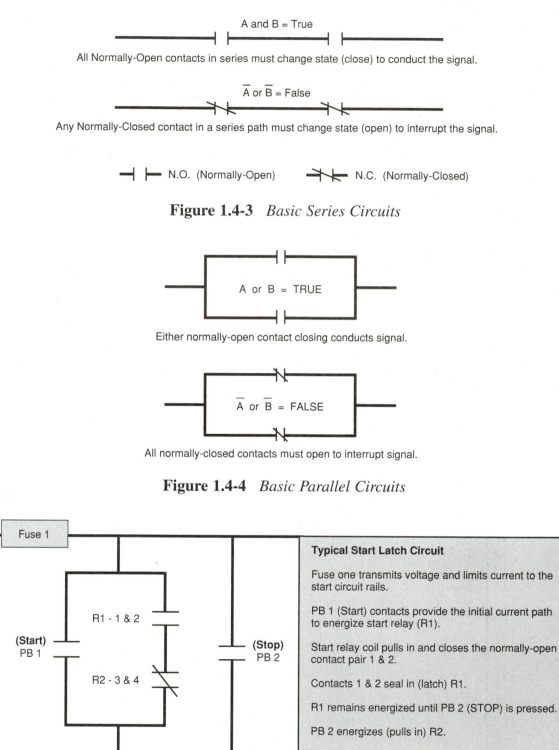

Fuse 1

R1 - 1 & 2

(Start)
PB 1

R2 - 3 & 4

(Stop)
PB 2

R1   R2

Typical Start Latch Circuit

Fuse one transmits voltage and limits current to the start circuit rails.

PB 1 (Start) contacts provide the initial current path to energize start relay (R1).

Start relay coil pulls in and closes the normally-open contact pair 1 & 2.

Contacts 1 & 2 seal in (latch) R1.

R1 remains energized until PB 2 (STOP) is pressed.

PB 2 energizes (pulls in) R2.

R2 normally-closed contacts, 3 & 4, open.

Current flow through R1 is interrupted.

R1 coil drops out forcing R1 - 1 & 2 open.

Figure 1.4-5 *Basic One Line Elementary Drawing*

1.5 Piping and Instrument Diagrams

Piping and instrument diagrams (P&IDs) are illustrations that display the actual hardware components associated with the flow path of the process. Interpreting P&ID's is much like reading a road map. The process media progresses through piping as it is transferred from place to place. Process control component symbols are affixed at each representative monitoring point and all processing devices are displayed. A multitude of symbols is used to diagram control loops. The first print in each set of drawings or prints usually contains a legend depicting the symbols used and their meaning. The American National Standards Institute and The Instrument Society of America publish the complete industry standard for instrument symbols and lettering abbreviations in the controlled document ANSI/ISA-S5.1.

Most instruments are represented as circles. The signal paths between associated instruments are depicted by various types of lines, Figure 1.5-1, and the instrument type is identified by an instrument function abbreviation, as listed in Figure 1.5-2. A numerical grouping hierarchy is also used to assist in clarification of component grouping and identification. Each instrument symbol should contain the function abbreviation, the process identification abbreviation, and the process control loop number, Figure 1.5-3.

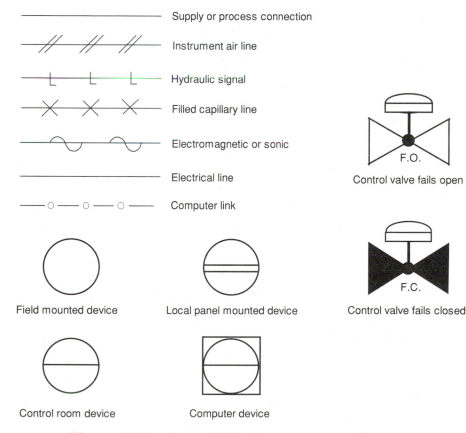

Figure 1.5-1 *Basic P&ID Instrument Symbols*

Letter	First Order Use	Succeeding Order Use
A	analysis	alarm
B	burner. flame	
C	user choice	controller
D	user choice	differential
E	voltage	element
F	flow	
G	user choice	gauge
H	hand	high
I	current	indicator
J	power	
K	time	control station
L	level	low
P	pressure	undefined
Q	quantity	totalize
R	radiation	recorder
S	speed/freq.	switch
T	temperature	transmitter
U	multivariable	multifunction
V	vibration	valve
W	weight	well
Y	event/state	solenoid/coil
Z	position	

Figure 1.5-2 *Instrument Function Abbreviations*

Figure 1.5-4 demonstrates all instruments monitoring and controlling a common variable sharing the assigned loop number.

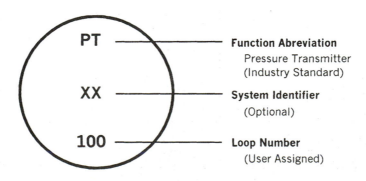

Figure 1.5-3 *Instrument Symbol Identification*

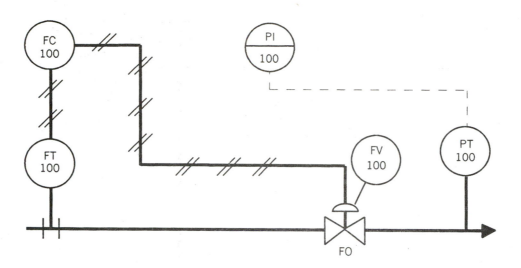

Figure 1.5-4 *Piping and Instrument Diagram*

1.6 Control Schemes

Common control schemes are feedback, feed forward, and cascade. Control requirements of the process involved determine the type of control scheme employed. Selection parameters include the degree of precision required, the inherent stability or volatility of the process, and the economic, environmental, or safety ramifications resulting from a lack of proper control. Each type of control has positive and negative attributes. Simpler control schemes are less expensive to install and maintain, but may not provide sufficient control for

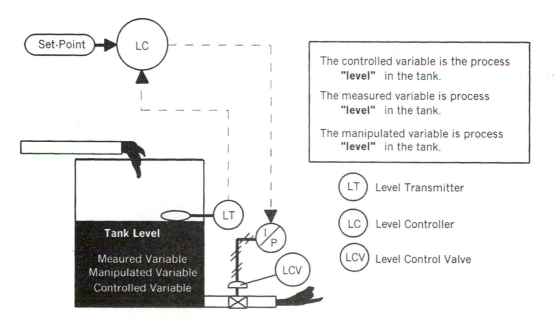

Figure 1.6-1 *Feedback Control*

complex or sensitive processes. Complex control schemes are more expensive to install and maintain, but provide extremely stable and precise control.

Feedback (reactive) control maintains the controlled variable at the desired value (set-point) by measuring the controlled variable, comparing the measured value to the set-point and manipulating the controlled variable until the set-point and measured value are the same. Figure 1.6-1 illustrates a typical example. Feedback control cannot respond until after the controlled variable changes.

Feed forward (pre-active) control measures one or more variables upstream of the controlled variable. A typical feed forward control loop, such as in Figure 1.6-2, manipulates these variables, prior to their introduction to the process, to minimize disturbance to the controlled variable. All possible effects of upstream disturbances in the measured variables are calculated to pre-determine the appropriate degree of manipulation required to prevent an upset in the controlled variable. If designed correctly, feed forward is a simple and stable control system. Any upsets or disturbances are measured and manipulated to prevent disruption of the controlled variable. Feed forward control does not directly measure or manipulate the controlled variable and is not capable of recognizing or responding to unanticipated deviations in the controlled variable.

Cascade (pre-active/re-active) control utilizes multiple controllers, with the output of one controller providing the set-point to another. A simple cascade loop is shown in Figure 1.6-3. The simplest concept of cascade control is an application that incorporates feed forward and feedback control in the same control loop. The output of the feed forward con-

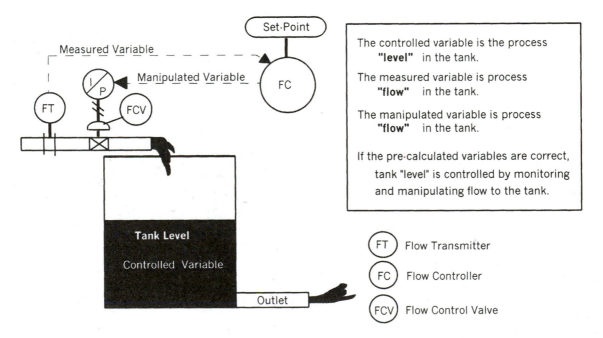

Figure 1.6-2 *Feed Forward Control*

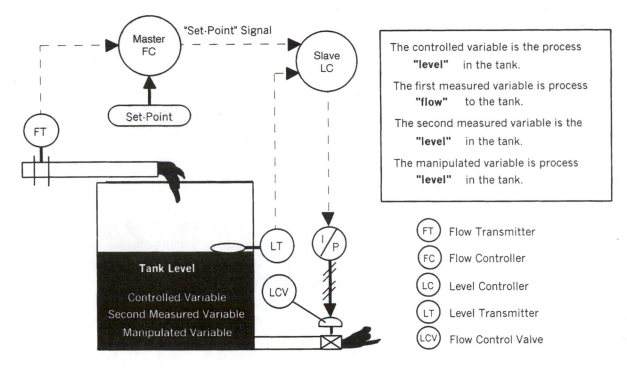

Figure 1.6-3 *Cascade Control*

troller establishes the set-point for the feedback controller. In the illustrated example, the feed forward section detects any disturbance upstream of the controlled variable. The disturbance forces an error signal in the flow controller. If flow increases, the controller outputs a new, lower set-point to the level controller. In response to the new, lower set-point, the level controller interprets the negative error as an increased level. The level controller transmits a valve open demand to the final control element. The feedback section then reacts to the increase in flow before the level actually changes. The combined pre-act/react response of cascade control scheme results in an extremely stable system. The additional cost and complexity of this scheme limits applications to inherently volatile or demanding processes.

There are certain processes that benefit from alternate control schemes. The room heater discussed earlier in this chapter cycles continuously between on and off as it tries to maintain temperature precisely at set-point. Differential gap control would reduce equipment wear and lower overall energy cost. Differential gap control attempts to maintain the process between a predetermined high and low value. This control demands the heater coil to energize when temperature drops significantly below set-point, and does not demand the coil to de-energize until temperature is at some value above set-point. The temperature range between the on and off demand is called dead band. While this is acceptable for a room temperature, some processes are quite intolerant of deviation and require more precise control. The automobile speed control unit discussed earlier in this chapter could benefit from multivariable control. Multivariable control accepts several different process variable inputs and computes an algorithm to produce the appropriate output demand. Monitoring engine vacuum and throttle position, in addition to speed, provides a more adaptive demand response. Ratio control is often used in mixing or chemical feed systems. Ratio control maintains a fixed ratio between two process variables. Complex processes often incorporate a master controller. A master control unit accepts inputs from several subordinate controllers and outputs a total system demand to each controller under its control.

1.7 Engineering Notation

Manufacturing and process industries are large-scale operations. As such, they produce massive quantities of product. On the other hand, quality, environmental, and efficiency concerns demand that extremely minute quantities are measured. Engineering notation simplifies the representation, communication, and documentation of this wide range of values. It annotates decimal place weights as groups of three. Figure 1.7-1 shows how each group of three places is assigned a specific prefix. A common transmission signal is "four to twenty milli-amps direct current." Engineering units allow representation of this value as $4 - 20$ mADC. Without engineering units, this value is $0.004 - 0.020$ ADC. A common analyzer value is micro mhos. 10 micro mho is an easier term to use than the equivalent 0.000010 mhos. Electrical generators commonly produce 35 megawatts. Without notation, this term is 35,000,000.00 watts.

Prefix	Symbol	Notation	Decimal Weight
Giga	G	$N \times 10^9$	N,000,000,000.00
Mega	M	$N \times 10^6$	N,000,000.00
Kilo	k	$N \times 10^3$	N,000.00
Centi	c	$N \times 10^{-2}$	0.0N
Milli	m	$N \times 10^{-3}$	0.00N
Micro	μ	$N \times 10^{-6}$	0.00000N
Nano	n	$N \times 10^{-9}$	0.0000000N

Figure 1.7-1 *Engineering Notation*

■ Chapter One Summary

● Process is a physical or chemical change of matter or conversion of energy.
● Control is the manipulation of variables influencing a process to achieve a desired result.
● Instrumentation is a device or group of related devices used to measure, monitor, and/or control a process. The purpose of process control instrumentation is to measure, monitor, and control a process.

- The four fundamental components of a control loop are the primary element, transmitter, controller, and final control element.
- Scales convert analog transmission signals into indications of actual process value.
- The linear or percent scale is graduated to indicate an equivalent percentage of process value in relation to percent of instrument signal.
- Square root scales are graduated to extract non-linear flow values from a linear instrument signal.
- The controlled variable is the parameter over which a desired or influenced result is anticipated.
- The measured variable is the parameter monitored by the primary element.
- The manipulated variable is the parameter directly altered by the final control element.
- The set-point is the desired control value.
- The primary element detects and converts energy from the process into a value suitable for measurement.
- The transmitter interprets the measurement from the primary element and transmits a standardized signal to the controller.
- The controller compares the actual value to the desired value (set-point) and applies preset algorithms to create a correction signal. An algorithm is a set of specific functions or instructions applied to a given variable.
- The final control element directly changes the value of the manipulated variable.
- Closed-loop control systems have feedback. Open-loop control systems operate without feedback.
- Binary logic evaluates conditions as a successive progression of TRUE/FALSE statements. TRUE is "1" and FALSE is "0."
- Interconnecting wiring, elementary, and one line drawings depict the function of individual measurement and control elements as a system of series and parallel lines.
- Series circuits provide a single path or route to conduct signal transmission or current flow. Parallel circuits provide multiple paths for signal transmission.
- Piping and instrument diagrams are illustrations that display the actual hardware components associated with the flow path of the process.
- The complete industry standard for instrument symbols and lettering abbreviations is established in the controlled document ANSI/ISA-S5.1.
- Feedback (reactive) control measures the controlled variable, compares the measured value to set-point, and manipulates the controlled variable until the set-point and measured value are the same. A deviation must exist before a corrective action occurs.
- Feed forward (pre-active) control measures and manipulates one or more variables upstream of the controlled variable, prior to their introduction to the process, to minimize disturbances to the controlled variable.
- Cascade (pre-active/reactive) control utilizes multiple controllers with the feed forward controller establishing the set-point for the feedback controller, and thereby changing the output of the feedback controller.

Chapter One Review Questions:

1. List the components of a control loop.

2. List the common control schemes.

3. Define the term process.

4. Define the term control.

5. Define the term instrumentation.

6. Define the term set-point.

7. Define the term measured variable.

8. Define the term manipulated variable.

9. Define the term controlled variable.

10. Define the term feedback.

11. State the purpose of process control instrumentation.

12 State the purpose of a primary element.

13. State the purpose of a transmitter.

14. State the purpose of a controller.

15. State the purpose of a final control element.

16. Illustrate a simple feedback control loop that includes a fail open level control valve on service water loop 100.

17. Illustrate a simple feed forward control loop that includes a fail closed level control valve on potable water loop 200.

18. Illustrate a simple cascade control loop that includes a fail open level control valve on potable water loop 100.

19. Develop the logic diagram to stop the engine if oil pressure and oil temperature are both high, or if vibration is high.

20. Develop the elementary diagram to latch "seal in" an engine over speed alarm relay.

Chapter 2

Primary Measured Variables

OBJECTIVES

Upon completion of Chapter Two the student will be able to:

- *List and identify the primary sensing elements used to detect pressure.*
- *State the hazards and precautions associated with pressure detectors.*
- *State the technique to determine operability of pressure detectors.*
- *List and identify the primary sensing elements used to detect temperature.*
- *State the hazards and precautions associated with temperature detectors.*
- *State the technique to determine operability of temperature detectors.*
- *List and identify the primary sensing elements used to detect level.*
- *State the hazards and precautions associated with level detectors.*
- *State the technique to determine operability of level detectors.*
- *List and identify the primary sensing elements used to detect flow.*
- *State the hazards and precautions associated with flow detectors.*
- *State the technique to determine operability of flow detectors.*
- *Identify common primary sensing element malfunctions.*
- *Define the following terms: primary sensing element, pressure, temperature, hydrostatic pressure, meniscus, static pressure, differential pressure, vena contracta, head correction, specific gravity, conduction, convection, radiation, and absolute zero.*
- *Differentiate between flow rate and total flow.*
- *Differentiate between mass and volume measurement as it pertains to flow.*

- *Convert units of pressure.*
- *Convert units of temperature.*
- *Calculate level sensor span and range.*
- *Calculate specific gravity level corrections.*

2.0 Introduction

Primary sensing elements detect and convert energy from the process into a value suitable for measurement. They are used to measure a multitude of process variables: flow, level, pressure, temperature, composition, etc. Process reaction energy is detected in a many physical forms. The common method is to monitor the actual value of concern. As this is not always possible or feasible, secondary parameters are sometimes monitored, and the critical value is inferred or extrapolated. This chapter investigates process variables and the primary devices used to detect and convert these variables into quantitative values.

2.1 Pressure

CAUTION **Process pressure devices may contain trapped pressure and/or hazardous materials. Obtain and review Material Safety Data Sheet for expected process media. Before disconnecting ensure proper drain and vent precautions are exercised. Always loosen device cautiously and observe for any evidence of trapped pressure.**

Pressure is defined as force exerted per unit area, Figure 2.1-1. A seventeenth century mathematician named Pascal documented several fundamental principles that hold true for pressure. The first principle states that the pressure at any point in a fluid acts equally in all directions. The second

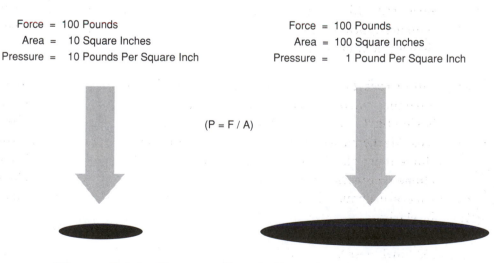

Force = 100 Pounds
Area = 10 Square Inches
Pressure = 10 Pounds Per Square Inch

Force = 100 Pounds
Area = 100 Square Inches
Pressure = 1 Pound Per Square Inch

$(P = F / A)$

Figure 2.1-1 *Pressure Equals Force Divided by Area*

Figure 2.1-2 *Pressures Equalize Across Entire System*

principle states that a change of pressure exerted at any point in a body of fluid at rest is transmitted undiminished to every point in the fluid. Pressure exhibits these principles as it always attempts to equalize throughout a system, Figure 2.1-2.

Scalar values of pressure require two terms. Practicality dictates common or accepted usages and conventions for combination of terms relating units of force to units of area. Common measured units of force are Newtons, pounds, and ounces. Common measured units of area are square meter, square centimeter, square millimeter, square feet, and square inches. Conventional statements of pressure combine these common terms. Examples of these combinations are newtons/meter^2 and pounds/inches^2. Since only like terms are manipulated and compared, conversion tables are frequently used. Common types of pressure sensing elements include manometers, bourdon tubes, and compression displacement elements.

Other parameters affect pressure values. Pressure in an enclosed system is dependent on temperature and volume. A balance exists between these three parameters and that balance always remains constant. Robert Boyle (1627 − 1691) and J.A.C. Charles (1746 − 1823) developed the basis of this relationship for gases. Mathematically stated pV/T = 1 (maintains unity). If temperature increases, pressure and/or volume must also increase. If temperature decreases, pressure and/or volume must also decrease. If volume increases, pressure and/or temperature must decrease. If volume decreases, pressure and/or temperature must increase. The following example assumes certain ideal conditions and is simplified for clarity.

Example: If a 1000^3 inch vessel is filled to capacity, pressurized to 100 psia, sealed, and stabilized at 373 degrees Kelvin, what is the maximum temperature allowable if the vessel burst at 300 psia?

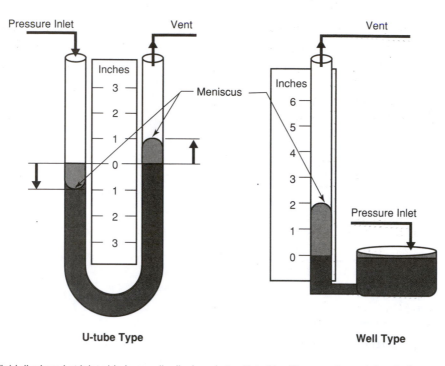

Fluid displaced at inlet side is equally displaced at outlet side. Pressure is read directly from scale.
Value is sum of total displacement. One inch plus one inch equals two inches pressure.

Figure 2.1-3 *Manometers*

> NOTE *Temperature is in Kelvin and pressures are absolute.*

Units of absolute pressure and volume are the same for both statements.

Solution: The initial condition is 100 psia × 1000^3/373 degrees Kelvin = 1

The final condition is stated as 300 psia × 1000^3/N degrees = 1

Since the relationships MUST remain constant, the initial condition is stated equal to the final condition.

If 100 × 1000/373 = 268, then 300 × 1000/N = 268, therefore N = 1119 degrees Kelvin.

Manometers are the simplest types of pressure detector. Manometers detect and convert a change in pressure into a linear vertical displacement of a known liquid. Evangelista Torricelli, who succeeded Galileo as professor of mathematics at the University of Florence, Italy, developed the first documented manometer. A manometer is a hollow transparent tube partially filled with pure water or mercury. As depicted in Figure 2.1-3, a ruled scale is affixed to the tube and the change of fluid level in the tube is read directly on the scale. The value indicated is equivalent to the amount of force required to lift the fluid the measured height. This type of pressure is called hydrostatic pressure. The fluid and scale used dictate the measurement units. If the fluid is water and the scale is inches, indicated values are read directly as "inches of water." If the fluid is mercury and the scale is graduated by millimeters the values are read as "millimeters of mercury." This is an extremely precise measurement if the liquid is of known pure concentration. The properties of pure water and the resistance of earth's gravity are extremely constant values.

NOTE *Install a pressure-limiting device upstream of the manometer. Pressure values greater than the manometer column height will expel the manometer fluid into the environment.*

CAUTION **Improper handling of mercury may result in personal injury or environmental contamination. Observe all applicable guidelines when storing, handling, and disposing of mercury.**

Manometer manufacturers provide colored fluids of precise specific gravity values for use in manometers. The colored fluids are easier to detect in the column than pure water.

Prior to recording data, ensure the manometer scale is adjusted to indicate zero with all pressure vented. Due to friction between the fluid and the inner glass cylinder wall, the fluid next to the wall resists displacement. The center of the fluid responds readily to the applied pressure and develops a concave or convex extrusion. The furthermost tip of this extrusion is called the meniscus.

NOTE *When reading the manometer scale observe the point where the tip of the meniscus is tangent to the center of the scale line.*

Bourdon tube elements, C type, spiral, and helix are elastic deformation elements. They convert fluid pressure into physical displacement. Used in some pneumatic controllers, these elements are most commonly used in local indication pressure gauges. Robert Hooke (1635 – 1703) developed several principles governing the design of modern pressure tubes. Key principles are stress, strain, and elasticity. Stress is the magnitude of a force applied to an object over the area of an object; strain is the measure of deformation caused by stress; and elasticity is the ability of a material to return to its original shape and size after an applied stress is removed. An object stressed beyond its elastic limit retains deformation after the stress is removed. Bourdon tubes are produced in C-type, spiral, and helix hollow metallic shapes. Bourdon C-type tubes resemble a question mark (?). Spiral tubes are wound flat and helix tubes resemble a coil spring. In each case, as increased pressure overcomes the spring tension of the tube it attempts to straighten or unwind. A metal link transfers this physical motion to a gear assembly. The gear assembly converts the linear motion of the tube into torque (a twisting motion). A pointer affixed to the gear indicates pressure on a calibrated circular scale. Manufactured elasticity and tensile characteristics of the tube determine its sensitivity and operating range. Bourdon elements should not operate above 75% full span for extended periods.

CAUTION **Due to inherent inaccuracy at zero, medium to high range pressure gauges are not reliable indicators of zero pressure. Use other means to ensure complete venting of a system.**

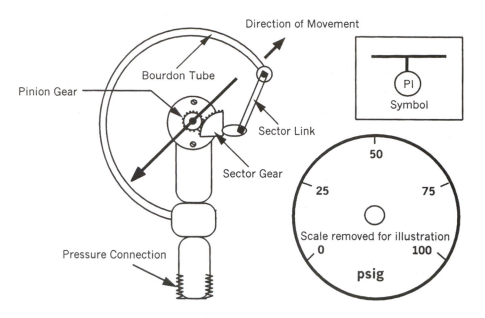

Figure 2.1-4 *Bourbon C-Type Tube Assembly*

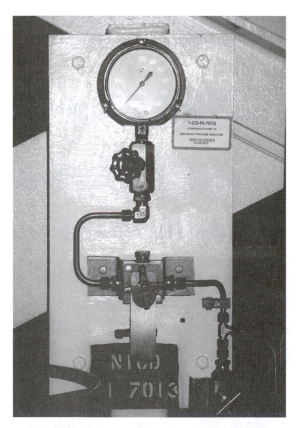

Figure 2.1-5 *Pressure Gauge with Adjustable Pulsation Damper*

Adjustable linkages, shown in Figure 2.1-4, permit fine-tuning.

Common compression displacement elements include the diaphragm, capsule, and bellows. Much like elastic elements, they convert changes in pressure into physical displacement. Elastic elements function as a spring; as pressure changes they deform and recover within their elastic limit. Compression elements are not designed to deform. Diaphragms act primarily to isolate and transfer changes in pressure from the external process media to an internal instrument fluid, Figure 2.1-6. Diaphragms are constructed as sensitive corrugated foil metallic disks. They are affixed to the outside of a metal well or cavity. The cavity is evacuated (all air removed) and filled with instrument oil. Once the cavity is sealed, any compressive displacement of the diaphragm reduces the volume of the cavity. As internal volume of the cavity is reduced, pressure within the cavity increases, and this increase is detected by additional instrumentation connected by capillary tubing to the diaphragm.

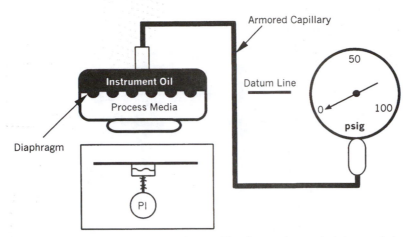

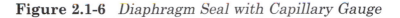

Filled capillary systems are elevation sensitive: if gauge is mounted above or below diaphragm seal elevation reference (datum line) zero compensation is required.

Figure 2.1-6 *Diaphragm Seal with Capillary Gauge*

> **NOTE** *Pressure devices attached to a diaphragm by capillary tubing are elevation sensitive. Adjustment is required to compensate for any internal hydrostatic head introduced through mounting locations.*

Capsules are constructed and operate similarly to diaphragms, however, provide more than isolation and transfer functions. Capsules have an electronic device or mechanical force beam internal to the capsule that converts physical displacement of the foil (compression) into an electronic or physical displacement suitable for monitoring by a transmitter. Figure 2.1-7 shows the arrangement of a differential pressure capsule whose output is the pressure difference between two points.

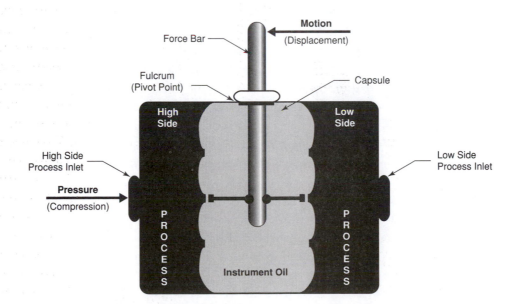

Figure 2.1-7 *Differential Pressure Capsule*

Flapper and nozzle convert bellows movement into a change of backpressure.

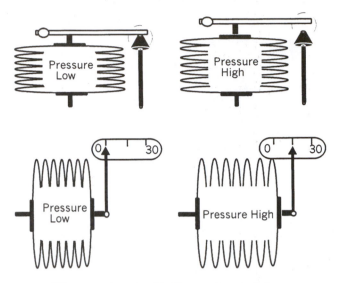

Figure 2.1-8 *Bellows Assembly*

Operability Assessment: Filled systems must maintain their hermetic seal to function properly. Any loss of fluid results in free space inside the capsule. Gentle tapping with a small smooth device that produces a hollow sound often reveals a loss of fluid and therefore a faulty component.

Bellows are constructed as hollow metallic corrugated cylinders. They convert pressure energy into direct linear motion. Bellows, shown in Figure 2.1-8, are typically constructed of brass, copper, or stainless steel. They are sensitive and fragile and measure only lower pressure ranges, typically limited to less than 30 psig. Some differential pressure gauges contain a set of opposing bellows attached to a torque tube assembly. The difference between the pressures applied to both bellows forces the bellows assembly to deflect. The torque tube converts bellows deflection into a rotational force. A pointer and adjustable link arm connected to the toque tube provides differential pressure indication. Bellows are tested by manually compressing the bellows assembly. While holding the assembly in the compressed position, seal the inlet tubing with your finger. The bellows should remain compressed until the seal is removed and the incoming air returns the bellows to its normal condition. Failure to maintain a seal indicates a leaking bellows. Several differential indicators use a filled bellows capsule assembly with a torque tube linkage drive. A range spring assembly determines the bellows' usable calibration range. Torque tube rotation requires re-centering if the range spring assembly is replaced.

Strain gauges and piezoelectric transducers are electronic pressure elements. Strain gauges convert physical displacement into a change of electrical resistance. The most common application of strain gauges is load cells. They are sometimes used in conjunction with compression displacers to convert motion into resistance. Strain gauges are constructed as a serpentine pattern of wire bonded to a backing. As the wire is stressed, the strain causes the wire to elongate. The resistance of a long, small-diameter wire is greater than the resistance of a short, large-diameter wire. Therefore as strain increases resistance also increases. Strain gauges are function tested using a multimeter in the ohms

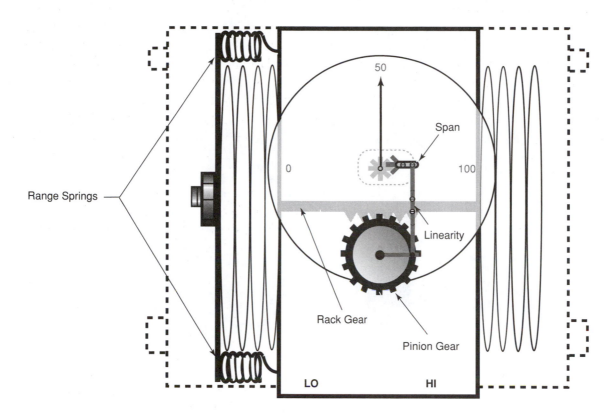

Figure 2.1-9 *Differential Pressure Bellows Unit*

position. An "open" reading indicates a bad strain gauge. Piezoelectric transducers convert physical displacement into electric voltage or electric voltage into displacement. The piezoelectric effect is a natural occurrence in many natural and artificial crystals. They are commonly used in conjunction with other devices.

Pressure sensing elements malfunction for a variety of reasons. The most common failures are attributed to over-ranging, rupture corrosion/erosion, plugging, loss of fill fluid, fatigue/brittleness, and excessive vibration. Figure 2.1-10 illustrates the uncharacteristic bulges or contours of an over-ranged element. Split seams and tears identify ruptures. Melting, pitting, and scarring are evidence of corrosion and erosion. Plugged instruments exhibit a total lack of response to system changes. Loss of fill fluid, Figure 2.1-11, is detected as a lack of response and a hollow or empty sound when the diaphragm is gently tapped. Fatigue and brittleness is detected as a loss of accuracy and lack of repeatability.

Pressure is measured in many different ways. Different units of force and area, as well as different points of reference are used. Typically, the magnitude of the variable measured determines the units and reference point used. To avoid confusion units are always identified with the scalar value. Pounds per square inch (psi) are used to identify larger magnitudes of measure. Pressures of lesser magnitude are identified as inches of water ("H$_2$O) or Inches of mercury ("Hg). Units for both gauge and absolute pressure are pounds per square inch (psi). The point of reference (0) for gauge pressure (pisg) is atmospheric pressure. Figure 2.1-12 demonstrates the point of reference (0) for absolute pressure

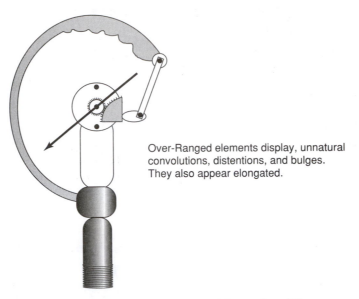

Figure 2.1-10 *Over-Ranged Bourdon Element*

(psia) is absolute zero. Pressures less than atmospheric are identified as inches of mercury vacuum ("Hg VAC) or millimeters mercury (mm Hg) absolute.

Conversion Factors:

1 pound per square inch (psi)

Equals   27.7   inches of water ("H$_2$O)

2.036 inches of mercury ("Hg)

51.71   millimeters of mercury (mm Hg)

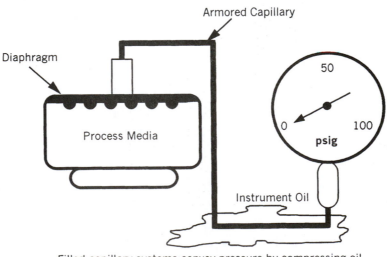

Figure 2.1-11 *Diaphragm Seal with Leak at Capillary*

Unit analysis is a method of conversion that assists in the preservation of correct value identities. This procedure incorporates the basic identity law; any number divided by itself is equal to one. The values of conversion are established as statements of ratio equalities. Numeric values are manipulated normally, but the unit identifiers are manipulated as if they were exponents.

Example: to convert 10 feet into inches determine the correct conversion factor:

1 foot is equal to 12 inches, therefore: 1 foot/12 inches = 1

State the unknown as a ratio.   10 feet/N = 1

Establish the ratios as an equality.   1 foot/12 inches = 10 feet/N

Solve for value N.   N × 1 foot = 10 feet × 12 inches

                                                                   N = (10 feet × 12 inches)/1 foot

Numerically   N = 120/1 = 120

Solve for Units N.   units N = feet × inches/feet

Restated   units = (feet/1) × (1/feet) × (inches/1)

Reduce units.   N = inches

Combined, the solution is 120 inches.

Example: Convert 10 inches of water pressure into millimeters of mercury.

27.7 inches of water is equal to 51.71 millimeter of mercury, therefore:

27.7 H_2O"/51.71 mm Hg = 1

State the unknown as a ratio.   10 H_2O"/N = 1

Establish ratios as an equality.   27.7 H_2O"/51.71 mm Hg = 10 H_2O"/N

Solve for value N.   N × 27.7 H_2O" = 10 H_2O" × 51.71 mm Hg

                                                   N = (10 H_2O" × 51.71 mm Hg)/27.7 H_2O"

Numerically   N = 517.1/27.7 = 18.67

Solve for units N.   units N = H_2O" × mm HG/H_2O"

Restated   units N = (H_2O"/1) × (1/H_2O") X (mm Hg/1)

Reduce   units N = mm Hg

Combined, the solution is 18.67 mm Hg.

Pressure measuring instruments are often exposed to vibration, pulsation, and corrosion. Pressure snubbers, pigtails, and gauge protectors, as shown in Figure 2.1-13, are often installed between the pressure device and severe processes. Pressure snubbers dampen pressure spikes and pulsations. Setscrews with small orifices drilled through them are often supplied with new gauges. The setscrew is installed into the threaded attachment portion of the Bourdon tube. Special snubber devices that resemble pipe fittings are also used. They are threaded onto the gauge connection between the gauge and process piping. Porous metallic filters or precision piston plungers inside the snubber prevent rapid pressure changes from damaging the gauge. Snubbers effectively dampen damaging pulsation to the gauge, but response time is attenuated, and they are susceptible to plugging. Pigtails are installed between the gauge and process piping to protect the gauge from steam and other harmful vapors. A section of pipe or tubing is formed into a loop. The loop collects and traps condensate to form a seal. The condensate isolates the gauge from extreme temperatures. Gauge protectors are installed on gauges used in extremely corrosive processes. The gauge protector is a diaphragm seal constructed of corrosion resistant material. A special evacuation and fill process is used to connect the gauge

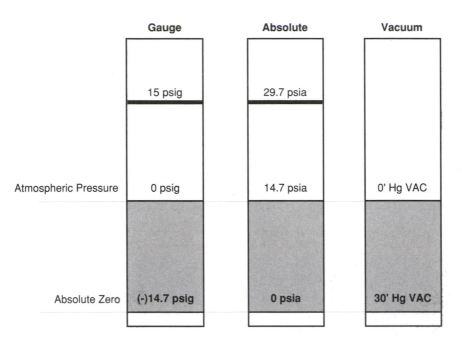

Figure 2.1-12 *Gauge Pressure/Absolute/and Vacuum*

to the gauge protector. Pressure gauges attached to diaphragm seals with capillary tubing are sensitive to mounting conditions. Mount the gauge and diaphragm at the same elevation, or adjust the gauge zero to compensate for the difference.

> **NOTE** *Disconnecting the gauge from the diaphragm protector will result in a loss of fill fluid and renders the gauge inoperable.*

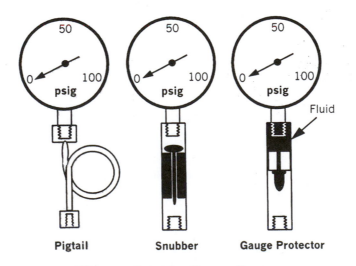

Figure 2.1-13 *Gauge Protectors*

2.2 Temperature

 Temperature processes are considered "high-energy" systems. They present a serious safety hazard. Use extreme caution and employ all reasonable precautions when performing work on "high-energy" systems.

Temperature is the basic measurement of thermal activity, or heat. Heat is the energy that flows between two bodies because of the temperature difference between them. Just as pressure attempts to reach equilibrium within a system, so too does heat. Heat transfers from one object to the next through three distinct processes. These processes, depicted in Figure 2.2-1, are conduction, convection, and radiation.

Conduction requires direct physical contact. Conduction is characterized as a transfer of heat energy throughout a body, without a transfer of substance through the body. Heat energy conducts as it energizes a continuous chain of molecules one after another. Conduction continues until heat energy among all adjacent molecules is equalized. An example of conduction occurs as an iron poker is placed among glowing fireplace embers. If the poker remains in the embers for a sufficient amount of time, heat energy at the fire end will transmit through the shaft to the handle.

Convection occurs as the actual substance containing the heat energy combines, or mixes with another substance. The resulting temperature is again equilibrium of the combined substances. Convection is a common form of heat transfer among fluids and gasses. Convection transfer occurs when a bit of cold water is added to cool a hot bath. Radiation transfer requires no substance to transmit energy. The heat energy is in the form of electromagnetic waves. The sun warms the earth's surface through radiation transfer.

Two phenomena of heat important to process instrumentation systems are phase changes and thermal expansion. Any pure substance can exist in one of three states: solid, liquid, or vapor.

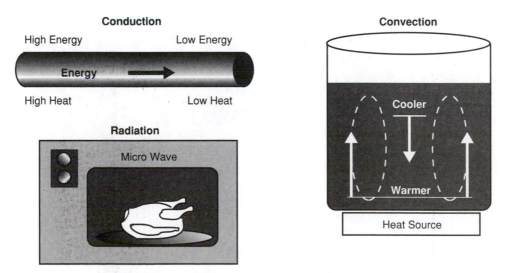

Figure 2.2-1 *Heat Transfer Methods*

All substances assume their physical state based on specific combinations of heat and pressure. The amount of energy absorbed by a substance during a phase transition or change of physical state is referred to as latent heat. Water in an automobile radiator maintains a liquid state at temperatures above its atmospheric boiling point. The water remains liquid if the radiator cap stays on and keeps the coolant pressurized. If the cap is removed while the temperature is above 212 degrees F, some of the water instantly changes from a liquid state into a vapor and flashes to steam. As a solid substance is heated, it eventually reaches its melting point. Additional heat energy cannot increase the temperature of a solid above this point. It does however force a phase change of the object from a solid to a liquid. Once the substance becomes liquid, it once again increases in temperature in response to heat energy. The temperature continues to rise until the liquid attains its boiling point. At this point, the temperature of the liquid does not increase. The liquid does however transition to a vapor.

The second phenomenon, thermal expansion, affects each substance to a different degree. Increasing temperatures generate molecular excitation (increased vibration). These vibrations cause substances to expand and become less dense. Solids become more pliable, liquids increase in volume, and gasses become lighter. Bi-metallic thermometers and other instruments incorporate thermal expansion as a means of determining temperature. Extended thermal cycling of certain materials induces stress from repeated expansion and contraction.

The two basic classifications of temperature sensing elements are mechanical and electronic. Both physical and electronic sensors often use thermo-wells. Thermo-wells are heavy walled metallic rods bored internally to accept the temperature probe, Figure 2.2-2. They provide process isolation, protecting the probe from corrosive and erosive processes and providing the ability to remove and replace probes without breaching the system.

Mechanical temperature elements include the bi-metallic, vapor pressure, and fluid pressure types. Bi-metallic elements are two metals with different thermal coefficients of expansion bonded together. As heat is applied, the difference in linear expansion causes the element to distort or bend.

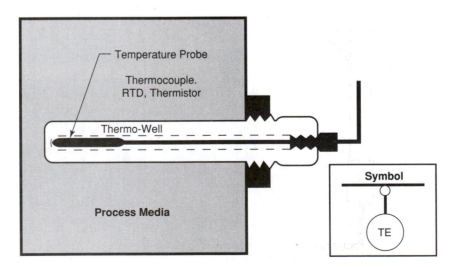

Figure 2.2-2 *Thermo-Well with Probe Installed*

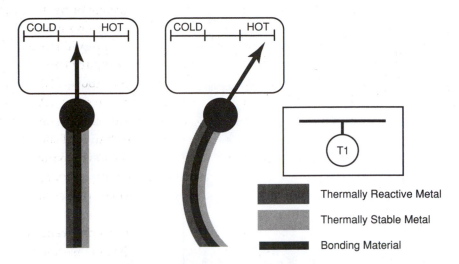

Figure 2.2-3 *Metalic Temperature Detector*

This distortion is converted into angular displacement and indicated on a scale by an attached pointer, Figure 2.2-3. The bonded bi-metallic is often wound into a spiral or helix to increase pointer travel and sensitivity. Filled thermal systems, Figure 2.2-4; adhere to the basic equilibrium theories of pressure, volume, and temperature. They convert temperature changes into equivalent pressure changes. Vapor pressure elements work on the phase transition principle. A filled bulb is capillary attached to a bourdon, spiral, or helix pressure element. The fluid in the bulb transitions to vapor as heat is applied. This gas or vapor pressurizes the enclosed system and the change in pressure is detected and indicated by the pressure element. Fluid temperature elements are constructed similar to vapor pressure elements. They are filled with fluid that expands at a known rate when heat is applied. The change in internal pressure is detected and indicated by a Bourdon, spiral, or helix pressure element.

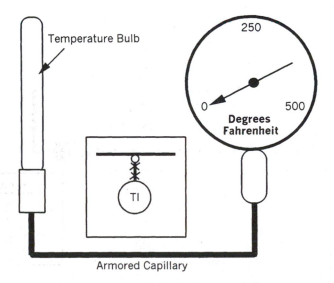

Figure 2.2-4 *Filled Thermal System*

ANSI Type	Color Code		Composition		Typical Range Degrees C
	(+)	(-)	Positive Element	Negative Element	
B	Gray	Red	Platinum (20% Rhodium)	Platinum (6% Rhodium)	0 – 1820
E	Purple	Red	Chromel	Constantan	-270 – 1000
J	White	Red	Iron	Constantan	-210 – 760
K	Yellow	Red	Chromel	Alumel	-270 – 1372
R	Black	Red	Platinum (13% Rhodium)	Platinum	-50 – 1768
S	Black	Red	Platinum (10% Rhodium)	Platinum	-50 – 1768
T	Blue	Red	Copper	Constantan	-270 – 400

Figure 2.2-5 *Thermocouple Color Code*

The primary electronic temperature-sensing elements are the thermocouple, RTD, and thermistor. Pyrometers are used when direct contact is not possible. The thermocouple is constructed of two specially selected dissimilar metals. In 1821, Thomas Seebeck discovered that when two wires of different metal were joined and heated on one end an electric current flowed through the wires. When the wires were disconnected on the unheated junction a voltage was present. Modern thermocouples are used with *electronic bridge circuits to increase flexibility and stability. Thermocouples have a proportional positive temperature coefficient. A temperature increase produces an equivalent voltage increase. Different metallic combinations are used to cover a wide range of temperatures and sensitivities. Figure 2.2-5 lists several types of thermocouples and temperature ranges. Resistance temperature detectors, RTDs, are constructed as a spiral or helix wound wire or a metallic film of pure metal. As the metal is heated the resistance increases, see Figure 2.2-6. A bridge circuit detects the change of resistance. Like the thermocouple, the RTD exhibits a proportionately positive output. Thermistors are thermally responsive resistors. They are man-made semiconductor devices and produce a negative exponential output response. Transistor or operational amplifier circuitry typically detects the change in resistance.

Mechanical temperature sensors are prone to zero drift errors over time. Due to the simplicity of design, they generally do or do not respond to temperature changes. Malfunction generally dictates replacement. Electronic sensors are fairly reliable. Malfunction of electronic sensors, other than physical damage, is usually a failure of associated electronic circuitry.

Operability Assessment: Test thermocouples with a DMM in DC volts position and compare value to a response chart. Test RTDs with a DMM in the ohms position and compare value to a response chart.

Electronic bridge circuits detailed in Chapter 3.3

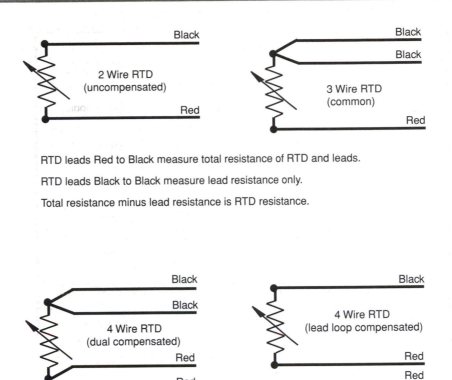

RTD leads Red to Black measure total resistance of RTD and leads.

RTD leads Black to Black measure lead resistance only.

Total resistance minus lead resistance is RTD resistance.

Red to Black should read 100 Ohms plus lead resistance at 0 degrees C.

Lead resistance varies with conductor resistivity, conductor length, and temperature.

Lead resistance for most field installations is typically low (2 – 10 Ohms).

Figure 2.2-6 *RTD Lead Configuration*

Similar to pressure, there are several units of measure used to identify temperature. The common units are degrees Celsius and Fahrenheit. Kelvin and Rankin are used for absolute measurement. Figure 2.2-7 identifies the three points of reference that assist in the comparison of scales. The ultimate point of comparison is absolute zero, the point totally devoid of all heat energy where molecular motion stops. At this point the Kelvin and the Rankin scale indicate 0°, the Celsius scale -273°, and the Fahrenheit scale -459.7°. The other two points are the transition points of pure water at sea level (one atmosphere). The freezing point of water is 32° Fahrenheit and 0° Celsius. The boiling point of water is 212° Fahrenheit and 100° Celsius. Between the freezing point and boiling point of water, a change of 1.8 degrees Fahrenheit is equivalent to a change of 1.0 degree Celsius. Fahrenheit and Rankin indicate the same amount of temperature change per degree. The primary difference is the point of reference; the Rankin scale begins at absolute zero. Celsius and Kelvin also indicate the same amount of temperature change per degree. The primary difference is again the point of reference; the Kelvin scale begins at absolute zero. The conversions for Fahrenheit and Celsius are °F = (1.8)(°C) + 32 and °C = (°F - 32)/1.8.

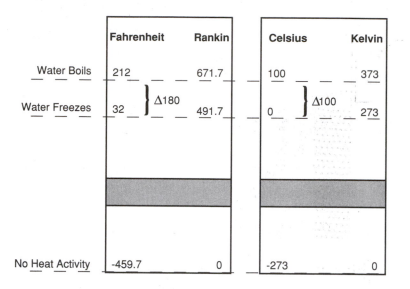

Figure 2.2-7 *Temperature Scale Comparison*

2.3 Level

Pressure, temperature, elevation, and density of the media monitored all effect level indication. Liquid level, just as pressure and temperature, attempts to maintain equilibrium throughout a system. Density is defined as mass per unit volume. Temperature effects fluid density. Increasing temperatures decrease density as the volume increases but the mass remains the same. Since differences in elevation effect level indications, the point of reference (datum line) is important to consider in level measurement. The lower sensing tap is the reference point for "0" percent level, and the upper sensing tap is the reference point for "100" percent level. Calculations to correct or compensate for

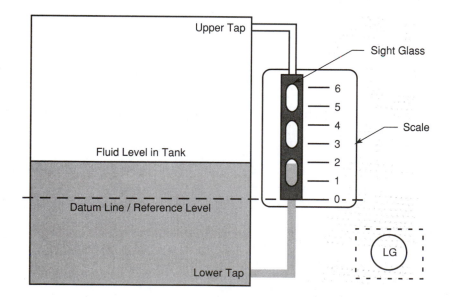

Figure 2.3-1 *Sight Glass Assembly*

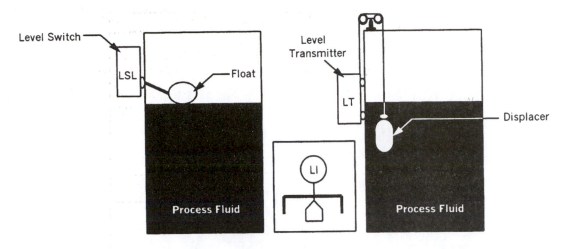

Figure 2.3-2 *Float and Displacer Comparison*

elevation differences are called head correction factors. Inches of water and feet of water are the standard measurement units for liquid head.

Figure 2.3-3 *Sump Pump with Float Switch*

Common types of level sensing elements include sight glasses, floats, displacers, and differential pressure transmitters. A sight glass is a section of transparent material, usually glass, piped in parallel to a vessel, tank, or column. As the fluid level in the vessel changes, the level in the sight follows. In Figure 2.3-1, the tank level is read directly from a scale etched or attached to the glass tube.

Displacers and floats are sealed balls or cylinders of known volume and density. They both adhere to the Archimedes' principle—the buoyancy of an immersed or partly immersed object is equal to the quantity of fluid it displaces. Floats are designed to remain at the surface of a fluid. Displacers are designed to attain a state of buoyant equilibrium at a known submerged level. Figure 2.3-2 illustrates the comparison of a float to a displacer. Each is attached to a tape, chain, or rod. A change in fluid level causes the float or displacer to rise or fall proportionally. The tape, chain, or rod transfers the motion to support instrumentation that converts motion into an electric, electronic, or pneumatic equivalent signal suitable for transmission. Figure 2.3-3 shows a sump pump and motor with a float control.

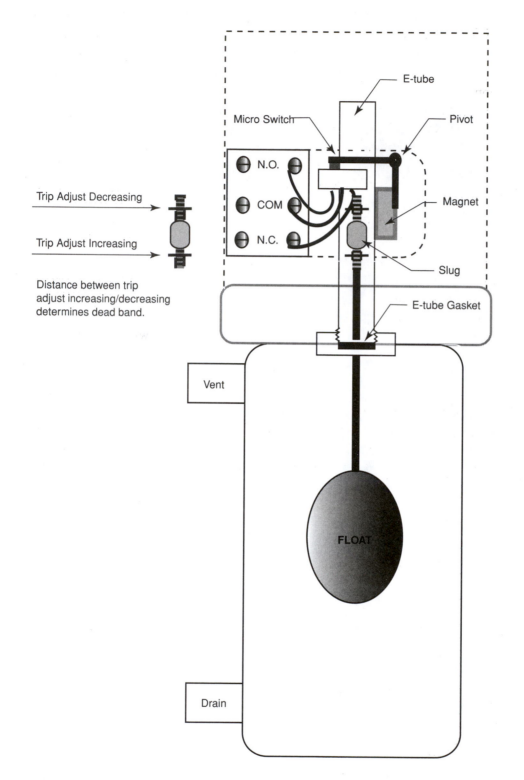

Figure 2.3-4 *E-Tube Level Switch*

Operability Assessment: Floats and displacers generally exhibit hard failures, they either float or they don't. If practical, vary the level of the fluid measured and observe the float or displacer. It is often possible to manually simulate the float movement and determine the operability of the detection device. A sight glass or level gauge is often used to verify and compare the actual level to the displacer indication.

Many float and displacer type level indicators use an E-tube to isolate the switch mechanism from the process. The float is directly attached to a threaded rod. A metallic sleeve, or slug, is mounted to the threaded rod. As the float level changes, the rod changes the position of the slug within the E-tube. A magnetically operated micro switch assembly detects the proximity of the metallic slug within the E-tube and changes state accordingly. Calibration is performed during initial installation by adjusting the location of the slug on the threaded rod. Once the E-tube is installed, no further adjustment is required, see Figure 2.3-4.

Another type of float/displacer device uses a torque arm or torsion beam to transmit displacer level to the internal mechanism. The displacer is attached to a twisting rod. Displacer tension causes the rod to flex and rotate slightly. A flapper nozzle assembly that is attached to the torque rod detects this rotation. Initial alignment requires the flapper nozzle and beam balanced with the displacer at midlevel position.

Differential pressure transmitters are often used to detect level. Pressure sensitive level instruments are installed with upper and lower sensing taps. Any process pressure in the vessel monitored is applied equally to opposite sides of the instrument and is therefore negated (algebraically summed to zero). Level measurement instrumentation is only sensitive to changes in level between the upper and lower range of the instrument. A tank indicates empty if the fluid level in the tank is below the lower sensing tap and below the measurement range of the level instrument. As in Figure 2.3-5, zero range suppression compensation is sometimes required to correct the indicated value. Density is the ratio of a fluid's mass divided by its volume. Relative density (specific gravity) is the ratio of a fluid's

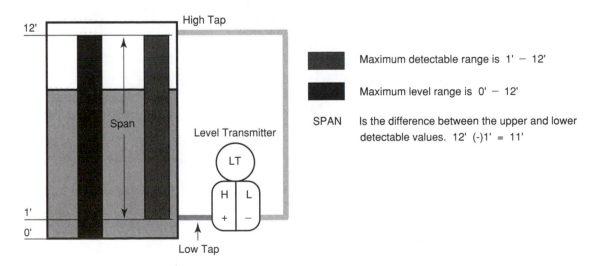

Figure 2.3-5 *Level Transmitter Detectable Range*

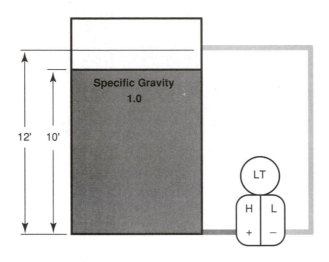

Maximum detectable range is 0 to 12' or = 0 − 144"

Liquid level is 10" x 12" = 120" H_2O

No density correction required if liquid is same density as water (s.g. 1.0)

Span is 144"   Measured value is 120"

Figure 2.3-6 *Range and Level Sg 1*

density divided by the density of water. Pure water has a specific gravity value of one. A correction factor is required for fluids that are more or less dense than water. Several comparisons are provid-

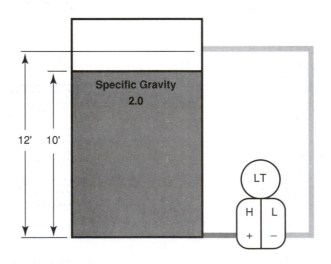

Maximum detectable range is 0 to 12' or = 0 − 144"

Liquid level is 10' x 12" = 120"

Density correction is 120" x 2 = 240" H_2O

Span is 288"   Measured value is 240"

Figure 2.3-7 *Range and Level Sg 2*

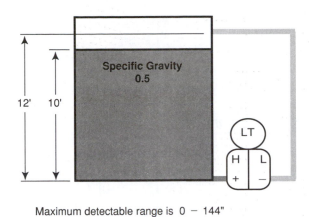

Maximum detectable range is 0 − 144"

Density correction is 144" x 0.5 = 72" H_2O

Liquid level is 10' x 12" = 120"

Density correction is 120" x 0.5 = 60" H_2O

Span is 72"   Measured value is 60"

Figure 2.3-8 *Range and Level Sg 0.5*

ed in Figures 2.3-6, 2.3-7, and 2.2-8. A fluid twice as dense as water has a specific gravity value of 2. A fluid half as dense as water has a specific gravity value of 0.5. The transmitters are piped directly to the vessel with an upper and lower sensing tap. Static pressure is equalized and only changes in level are detected as the algebraic difference between the total high side (positive) value and the total low side (negative) value. This is clarified by Figures 2.3-9 and 2.3-10. Upper tap sensing lines are

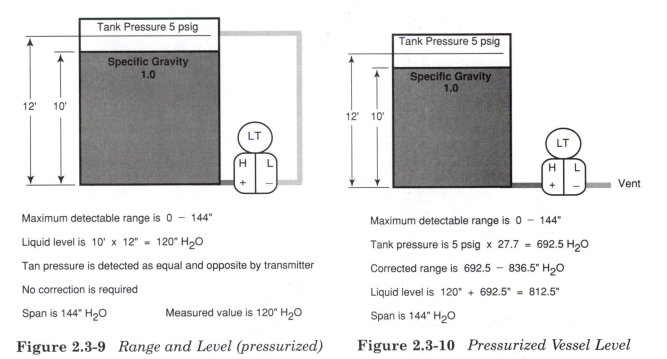

Maximum detectable range is 0 − 144"

Liquid level is 10' x 12" = 120" H_2O

Tan pressure is detected as equal and opposite by transmitter

No correction is required

Span is 144" H_2O Measured value is 120" H_2O

Figure 2.3-9 *Range and Level (pressurized)*

Maximum detectable range is 0 − 144"

Tank pressure is 5 psig x 27.7 = 692.5 H_2O

Corrected range is 692.5 − 836.5" H_2O

Liquid level is 120" + 692.5 = 812.5"

Span is 144" H_2O

Figure 2.3-10 *Pressurized Vessel Level*

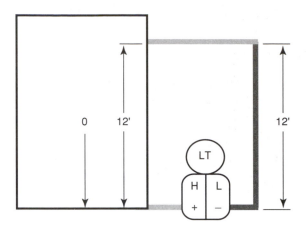

Maximum detectable range is 0 to 12' x 12" = 0 – 144"

Liquid level is 0' x 12" = 0" H_2O

Low side filled reference is always 12' x 12" = 144"

Calibration range is (-)144" to 0" H_2O Span remains 144" H_2O

Measured value is algabraic difference at transmitter 0" – 144" = (-)144" H_2O

Figure 2.3-11 *Vessel Level Empty*

sometimes backfilled with a reference fluid to prevent over range damage or reverse the operation of the sensor capsule. Specific applications are developed in Figures 2.3-11, 2.3-12, and 2.3-13. Most transmitters are equipped with vent plugs to assist in air removal.

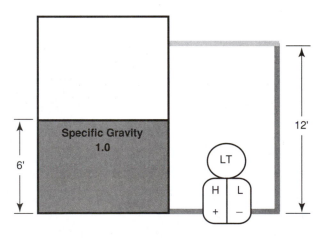

Maximum detectable range is 0 to 12' x 12" = 0 – 144"

Liquid level is 6' x 12" = 72" H_2O

Low side filled reference is always 12' x 12" = 144"

Calibration range is (-)144" to 0" H_2O Span remains 144" H_2O

Measured value is algabraic difference at transmitter 72" – 144" = (-)72" H_2O

Figure 2.3-12 *Vessel Mid-Level*

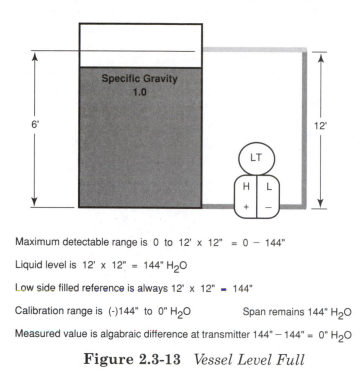

Maximum detectable range is 0 to 12' x 12" = 0 − 144"

Liquid level is 12' x 12" = 144" H_2O

Low side filled reference is always 12' x 12" = 144"

Calibration range is (-)144" to 0" H_2O Span remains 144" H_2O

Measured value is algabraic difference at transmitter 144" − 144" = 0" H_2O

Figure 2.3-13 *Vessel Level Full*

Determine the risk associated with the process media before attempting to vent the transmitter. Determine the direction of the vent hole and provide suitable catch containers as required. Aerated fluids appear opaque and exit the vent in erratic spurts. Non-aerated fluids retain their original appearance and exit the vent in a uniform manner. Continue venting until all air is removed. At no time, expose the transmitter to excess differential pressure.

Another level sensing application of differential transmitters is the bubbler tube. In Figure 2.3-14, tubing is mounted vertically inside the tank with the outlet very near the bottom. An air regulator applies a small fixed volume of air to the tube. Air pressure in the tube increases until it is equal to the hydrostatic pressure at the exit point of the tube. Air pressure in the tube remains equal to the hydrostatic backpressure as long as bubbles continue to exit the tube. A transmitter connected to the tube detects the amount of backpressure. Proper level indication is obtained when the air bubbles exit the tube one at a time with stable periodicity (evenly spaced). The flow rate is adjusted correctly when the flow meter indication sphere appears to bounce slightly within the glass cylinder. Excess restriction in the bubbler tube creates an artificially high reading. Leaks in the bubbler tube create artificially low readings.

Operational Assessment: If a slight increase in the rotameter flow rate increases the indication, the tube is probably plugged. If a slight decrease in the rotameter flow rate decreases the indication, the tube probably has a leak.

There are several electronic level instruments. The most common are capacitance probes, ultrasonic probes, and radiation transmitters. Capacitance probes are direct contact level sensors. In Figure 2.3-15, two parallel conductive rods extend vertically into the fluid. The rods are fed an

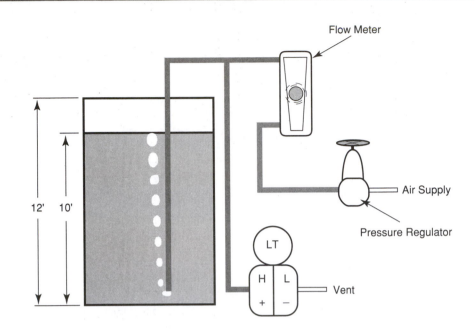

Figure 2.3-14 *Air Bubbler System*

excitation voltage. The capacitance between the rods is determined by the amount of charge (excitation), distance between the rods, and the dielectric strength of the material between the rods. Since the charge and distance are fixed, the only variable is dielectric strength. All fluids have much higher dielectric constant values than air. The level of fluid at maximum rod submersion creates maximum dielectric coupling and maximum capacitance. As the level of fluid in the vessel decreases, the dielectrically coupled area between the two rods decreases and capacitance drops.

Ultrasonic and radiation level detectors are non-contacting probes. Ultrasonic transmitters emit a pulsed frequency signal of known wave velocity from the top of the vessel into the perpendicular

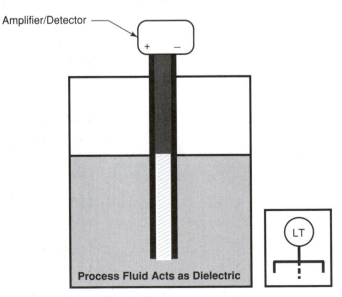

Figure 2.3-15 *Capacitance*

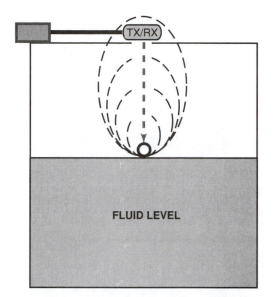

Figure 2.3-16 *Ultrasonic Level Detector*

surface of the fluid, Figure 2.3-16. To determine wave velocities multiply the signal wavelength by the signal frequency. The time delay between the transmission of the original pulse and detection of the reflected signal is proportional to the distance the pulse travels. The elapsed time represents the entire wave cycle, from the transmitter to the fluid and back to the receiver. The actual distance from the transmitter to the fluid is one half of a cycle. Since ultrasonic probes reference level from the top down, they actually detect the amount of fluid not in the vessel. Electronic circuits in the transmit/receive module perform the necessary calculations and conversions. The signal must approach the fluid surface perfectly perpendicular for maximum signal strength

Radiation transmitters are also non-contacting level detectors. They are two-piece units consisting of a radiation transmitter (source) and a corresponding radiation detector. Radiation travels relatively undiminished in free space. Any fluid present absorbs radiation at a rate specific to the properties of that fluid. The source and detector are placed at the same level directly across the tank from each other. The transmitter emits gamma waves at a fixed known rate. The difference between the energy transmitted and the energy received is representative of the amount of shielding or fluid between the transmitter and detector. High radiation levels indicate an empty vessel. Lower radiation levels indicate the presence of fluid between the transmitter and receiver. Since each transmitter/receiver pair only determines the presence or absence of fluid within a narrow range, several pairs are required to track significant level fluctuations, Figure 2.3-17. Although not considered immediately hazardous, low-level radiation sources do require special handling, usage, and disposal considerations.

Operability Assessment: Sight glass, float, and displacer sensing element malfunctions are usually leaks and sticking floats. Pressure element malfunctions include loss of reference leg, plugged sensing lines, loss of fill fluid, and improper bubbler adjustment. Capacitance probes are susceptible to corrosion, and galvanic erosion, and do not work with non-conductive fluids. Ultrasonic and radiation element problems are typically detector/transducer failures and misalignment. Ultrasonic

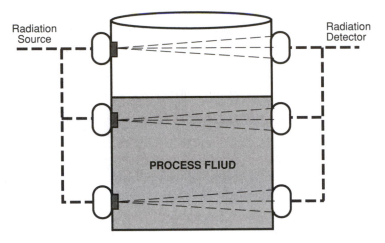

Figure 2.3-17 *Radiation*

signals must approach the fluid surface perfectly perpendicular for maximum signal strength. Radiation transmitter/detector pairs mounted perfectly diametric in the same plane generate maximum signal strength.

2.4 Flow

Flow rate measurement monitors the instantaneous volume of process media transferring per unit of time. Total flow is the accumulated volume transferred over a period of time. Flow rate terms include both units of volume (gallons, cubic feet, cubic centimeters) or mass (moles, pounds), and units of time (hours, minutes, seconds). Typical flow rate units are gallons per second and cubic feet per minute. Flow rate measurements of volume require compensations for physical and chemical factors. Pressure, density, viscosity, and velocity all affect flow measurement. Daniel Bernoulli (1700 – 1782) developed the principles of flow measurement in use today. Bernoulli's fundamental dynamics of flow states that there is an inverse relationship between pressure and velocity. If fluid velocity increases, pressure decreases and if velocity decreases, pressure increases.

Additional important concepts pertain to incompressible fluids. The volume, mass, and energy of a fluid in motion must maintain continuity. Simply stated, whatever enters one end of a pipe is equivalent to whatever exits the other. The relationship of flow rate, velocity, and differential pressure is simplified by the equation $Q = K/DP$. The principle statement is a derivation of Bernoulli's equation for ideal flow of a noncompressible fluid (liquid), $Q = K\,A_o\,\sqrt{2gh}$. Q is the flow rate (volume per unit of time), K is the efficiency factor (a correction factor obtained from charts or tables), A_o is the area of the orifice in square feet, g is the acceleration of gravity (32 feet/second/second), and h is the difference in pressure across the orifice. Compressible fluid (gas) calculations must include static pressure (the amount of compression). The formula for gas flow is $Q = K\,\sqrt{hp}$. Q is flow in cubic feet per hour, K is a conversion correction factor (obtained from tables), h is the difference in pressure across the orifice, and p is the upstream static pressure. Static pressure is the actual total system pressure. Differential pressure is the difference in pressure between the upstream

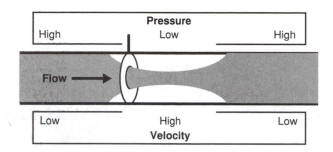

Figure 2.4-1 *Bernoulli Effect*

and downstream tap. Differential pressure increases with flow rates, but not linearly; as shown in the equation the differential pressure increases with the square of the flow rate.

Mechanical flow sensing elements are the rotameter, orifice plate, venturi, flow nozzle, annubar, Pitot tube, elbow tap, weir, and flume. The rotameter, orifice plate, venturi, flow nozzle, annubar, and Pitot tube actually detect pressure differential resulting from changes in velocity. The physical effect, depicted in Figure 2.4-1, resembles an hourglass and the point of maximum velocity and minimum pressure is called the vena contracta. Each of these detectors creates some restriction to flow.

Figure 2.4-2 *Flow Transmitter*

The restriction forces an increase in velocity and sensing taps located upstream and downstream of the restriction detect the resultant pressure differential.

Rotameters are used for local indication only. They are constructed of a ball or float located inside a tapered tube, as depicted in Figure 2.4-3. The ball restricts the fluid flow. As pressure increases, the ball rises in the tapered tube and allows increased flow through the tube. The equivalent flow rate values are read directly off the tube or an attached scale.

Orifice plates are the most common flow-sensing element, Figure 2.4-4. They are constructed as a metallic disk with a precision hole machined through the face. Concentric orifice plates have the orifice hole located in the center of the plate. Eccentric orifice plates have the hole located just below the center of the plate. Segmented orifice plates have a semicircular hole usually located just below the center of the plate. The type of orifice plate selected is determined by the flow characteristics of the product. Orifices are

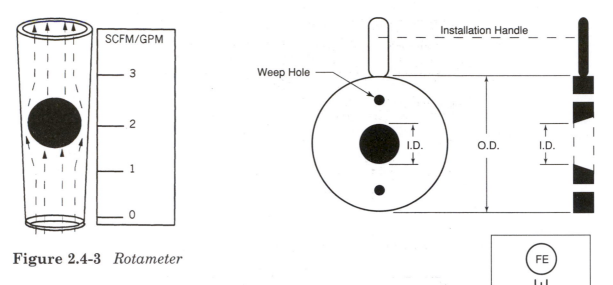

Figure 2.4-3 *Rotameter*

installed directly into the piping. Interior diameter, exterior diameter, and installation orientation are stamped or engraved on the orifice plate handle. Orifices are

Figure 2.4-4 *Sharp Edge Orifice Plate*

generally located at least ten pipe diameters downstream and five pipe diameters upstream of any turns, pumps, or valves. Pressure sensing taps are installed to monitor pressure drop across the orifice, Figure 2.4-5. Weep holes are sometimes provided in the orifice plate to prevent build-up in front of the plate. The orifice is installed with the sharp face upstream. Orifices should be installed at least ten pipe diameters downstream and five pipe diameters upstream of any flow disturbances.

The venturi and flow nozzle function under the same principles as orifice plates. A venturi is an hourglass shaped restriction in the piping, as demonstrated in Figure 2.4-6. It is less susceptible to plugging than an orifice and creates less restriction to flow. Flow nozzles are basically the upstream section of a venturi.

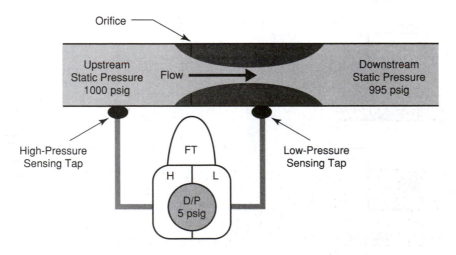

Figure 2.4-5 *Static Pressure*

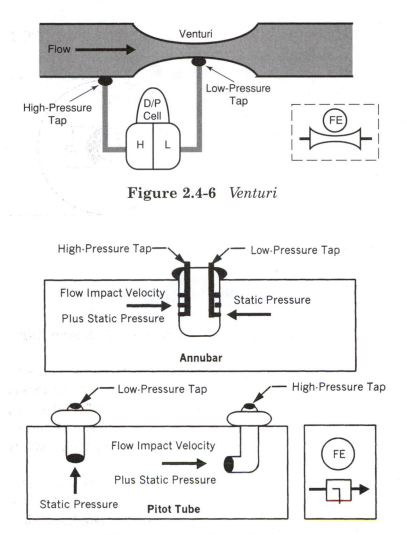

Figure 2.4-6 *Venturi*

Figure 2.4-7 *Pitot/Annubar*

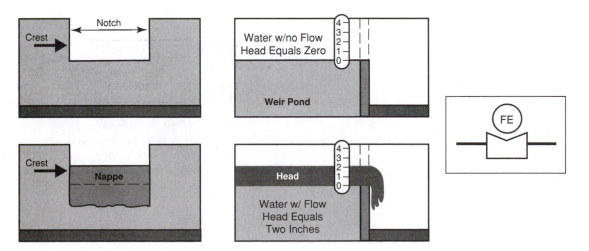

Figure 2.4-8 *Wier*

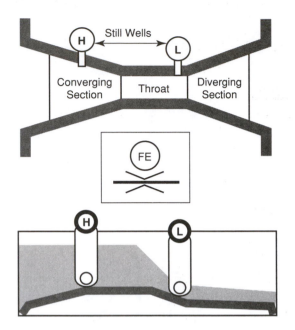

Figure 2.4-9 *Flume*

Annubars and Pitot tubes, shown in Figure 2.4-7, are specially shaped tubes with precision machined sensing vents. The tubes are inserted directly into the flow media perpendicular to the flow. Pressure sensing lines are attached directly to the tubes. As with all flow detection devices the high-pressure area is upstream of the restriction.

Elbow taps are mounted at the inside and outside of ninety-degree bends in the piping, and do not restrict flow. Centrifugal energy of the fluid in motion generates a high-pressure region on the outside of the bend. The difference in pressure is calculated into flow rate. Weirs and flumes are used to measure flow in open channels. Figure 2.4-8 depicts a typical weir. A weir is constructed as a box placed directly in the channel. The upstream area of the box serves to calm the fluid. The actual restriction is a precisely measured weir plate. The theory of measurement derives the cubic volume of fluid flow by multiplying the flow velocity in feet per minute by the square feet area of the water head at the weir notch. Actual flow figures are usually obtained from a special chart called a nomograph. In 1926, Ralph Parshall documented the use of the flume to determine the flow rate of agricultural water. Flumes are generally constructed of concrete. Figure 2.4-9 depicts a typical flume. The sides of the flume resemble a venturi in shape. The fluid flow is directed through a narrowed section of the channel; since the fluid is not enclosed in piping the restriction results in an increased level in the channel or gate. The change in level is read directly from scaled measuring rods in the fluid, detected by a float or displacer in attached still wells, or sensed as hydrostatic pressure by a pressure device.

Electronic instruments also measure flow. The ultrasonic transducer, turbine flow meter, and electromagnet flow meter detect flow velocity. The heated junction converts heat loss into mass flow. The ultrasonic transducer consists of a transmitter and a receiver. The transducers are located on opposite sides of the pipe, linearly offset to transmit diagonally across the flow. Process media assist

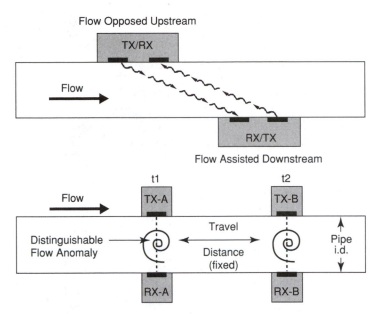

Flow Opposed Upstream

TX/RX

Flow

RX/TX

Flow Assisted Downstream

t1   t2

Flow   TX-A   TX-B

Distinguishable Flow Anomaly   Travel   Pipe i.d.

Distance (fixed)

RX-A   RX-B

t2 - t1 /d = Flow Velocity
Corrected volume of product per pipe area X flow velocity = flow rate

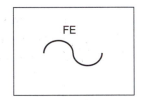

FE

Figure 2.4-10   *Ultrasonic Flow Detector*

ultrasonic transmission in the downstream direction and oppose transmission in the upstream direction, Figure 2.4-10. This aiding and abating is detected as a difference in propagation times. Increased flow decreases downstream delay times and increase upstream delay time.

The turbine meter shown in Figure 2.4-11, consists of a turbine or propeller mounted inside the pipe facing the flow. Fluid flow departs energy to the turbine causing the blades to spin at a rate directly

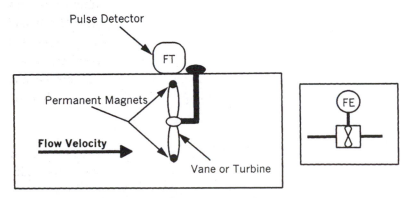

Pulse Detector

FT

Permanent Magnets

Flow Velocity

FE

Vane or Turbine

Figure 2.4-11   *Pulse Generator*

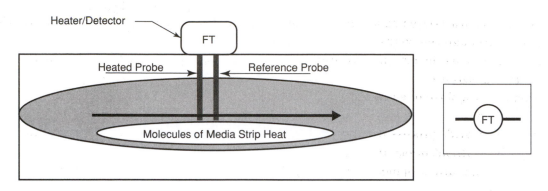

Figure 2.4-12 *Mass Flow Meter*

proportional to the fluid velocity. Magnets attached to the blades produce a pulse as they spin past a detector. The pulse frequency is electronically converted into flow velocity.

Heated junction flow detectors are molar specific. Since they detect the actual mass of flow, they automatically compensate for other variables such as pressure, density, and viscosity. The heated junction consists of an element or wire directly exposed to the process media. Electronic circuits drive current through the wire to hold it at a fixed temperature. Figure 2.4-12 demonstrates the flowing media stripping heat energy from the wire. The amount of current required to offset the heat loss is equivalent to the flow mass absorbing the heat. The changes in heater current are detected and processed by additional circuitry.

Operability Assessment: Plugging, corrosion, erosion, loss of reference or compensation, and improper installation are the most common flow sensing element malfunctions. Most flow restriction devices experience some corrosion, erosion, or fouling over time. Critical flows are often monitored simultaneously by several techniques and the resultant values are continually compared.

■ Chapter Two Summary

- Primary sensing elements detect and/or convert energy from the process into a value suitable for measurement.
- Pressure is defined as force exerted over a specific area.
- A change of pressure exerted at any point in a body of fluid at rest is transmitted undiminished to every point in the fluid. Pressure in an enclosed system is dependent on temperature and volume.
- The amount of force exerted by a column of fluid is called hydrostatic pressure.
- When reading the manometer scale observe the point where the tip of the meniscus is tangent to the center of the scale line.
- The point of reference for gauge pressure (psig) is atmospheric pressure. The point of reference for absolute pressure (psia) is absolute zero. Pressure less than atmospheric is also identified as inches of mercury vacuum ("Hg VAC).
- Temperature is the basic measurement of thermal activity, or heat.

- Conduction is characterized as a transfer of heat energy throughout a body, without a transfer of substance through the body.
- Convection occurs as the actual substance containing the heat energy combines, or mixes with another substance.
- Radiation is in the form of electromagnetic waves; transfer requires no substance to transmit energy.
- Thermo-wells provide process isolation, protecting the probe and providing the ability to replace probes without breaching the system.
- Mechanical temperature elements include the bi-metallic, vapor pressure, and fluid pressure types.
- Three primary electronic temperature-sensing elements are the thermocouple, RTD, and thermistor.
- Temperature units are degrees Celsius, Fahrenheit, Kelvin, and Rankin. Absolute zero is the point totally devoid of all heat energy where molecular motion stops.
- Pressure, temperature, elevation, and density of the media monitored all affect level indication.
- Liquid level attempts to maintain equilibrium throughout a system.
- Common types of level sensing elements include sight glasses, floats, displacers, and differential pressure transmitters.
- Differential pressure transmitters are installed with upper and lower taps to detect level.
- A correction factor, specific gravity, is required for some level instruments measuring fluids that are more or less dense than water. Pure water has a specific gravity value of one (1).
- Flow rate measurement monitors the instantaneous volume of process media transferring per unit of time. Total flow is the accumulated volume transferred over a period of time.
- Pressure, density, viscosity, and velocity all affect flow measurement.
- The point of maximum velocity and minimum pressure is called the vena contracta.
- Static pressure is actual total system pressure. Differential pressure is the difference in pressure across a restriction.
- Mechanical flow sensing elements are the rotameter, orifice plate, venturi, flow nozzle, Annubar, Pitot tube, elbow tap, weir, and flume.
- Weirs and flumes are used to measure flow in open channels.

Chapter Two Review Questions:

1. List and identify the primary sensing elements used to detect pressure.

2. List and identify the primary sensing elements used to detect temperature.

3. List and identify the primary sensing elements used to detect level.

4. List and identify the primary sensing elements used to detect flow.

5. Identify primary sensing element malfunctions.

6. Define the term primary sensing element.

7. Define the term hydrostatic pressure.

8. Explain the correct technique to observe a manometer reading.

9. Differentiate static pressure and differential pressure.

10. Explain the relationship between head correction and specific gravity.

11. Differentiate between flow rate and total flow.

12. Convert 10 psig into psia.

13. Convert 10 psia into "Hg VAC.

14. Convert 12 feet H_2O into psig.

15. Convert 277" H_2O into psig.

16. Convert 100 degrees Celsius into degrees Fahrenheit.

17. Convert 32 degrees Fahrenheit into degrees Celsius.

18. Calculate level sensor span and range in "H_2O for a D/P transmitter attached to a vessel of pure water with the lower tap installed three feet from the tank bottom and the upper tap installed fifteen feet from the tank bottom.

19. Perform the calculation for the previous system (18) assuming the process fluid is twice as dense as water.

Chapter 3

Standards and Methods

- *Analyze and interpret a schematic diagram to determine adjustment locations required to eliminate offset error and multiplication error.*
- *Analyze and interpret a mechanical lever mechanism to determine adjustment locations required to eliminate offset error, multiplication error, and linearity error.*
- *Analyze and interpret a force balance beam assembly to determine adjustment locations required to eliminate offset error and multiplication error.*

3.0 Introduction

Early Egyptian, Roman, and Hebrew civilizations used a unit of measure called the cubit. The cubit was defined as the distance between the tip of the middle finger and the elbow of a man's arm. King Henry I (1068 – 1135) decreed the yard as the distance from the tip of his outreached fingers to his nose. Shortly after the French Revolution, the foundation of our modern measures was laid. On June 22, 1799 two platinum bars, one designated the "meter" and the other designated the "kilogram," were placed in evacuated glass cases in the Archives de la Republic in Paris, France. Duplicates of these artifacts were produced and distributed to standard laboratories around the world. As technology developed, increasingly repeatable standards were developed. In 1960, the meter was redefined in relation to spectral wavelengths of the krypton-86 atom, a known physical constant, and became an intrinsic standard. The current meter is defined as the distance traveled by light in a vacuum over a time interval of 1/299 792 458 of a second. Standards based on an artifact require access to the artifact, or a duplicate of the artifact. Intrinsic standards are based on a known physical constant and are reproducible by each standards laboratory. The standards custodian for the United States is the National Institute of Standards and Technology, formerly known as the National Bureau of Standards.

3.1 Traceability and Documentation

Calibration is the process of comparing the outputs of a device to a known traceable standard, and/or adjusting the outputs of a device to replicate the standard within a specified tolerance. A calibration cycle consists of multiple point comparisons to a standard over the entire range of the device from both increasing and decreasing directions. Trace ability is the documentation process that validates the lineage of a measurement through an unbroken chain of standards to a national standards laboratory. A calibration report is a graph or table that contains the comparison data between a standard and the device under calibration.

This report is often called a calibration data sheet. The calibration report should include the serial number, calibration date, and calibration expiration date of the standard used. It should also contain the desired value of all data points monitored, with minimum and maximum acceptable values see Figure 3.1-1. The report should provide data blanks to record as found and as left readings. The report may contain additional notes, or information required to standardize the calibration method

```
MEASURING AND TEST EQUIPMENT I.D. TAG
MANUFACTURE _____
MODEL # _____  SERIAL #_____
CALIBRATED BY_____  DATE _____
CALIBRATION DUE DATE _____
TEST EQUIPMENT USED _____
TEST EQUIPMENT DUE DATE _____
```

SCALE DATA SHEET   **FW-PI-101**

DESCRIPTION: **FEEDWATER PRESSURE INDICATOR**

TOLERANCE: +/- 1/2%   **HEAD CORRECTION: NONE**

DATA SHEET REVISION #_____   APPROVAL DATE_____

Req	Min	As Found	Max	Dev.	Min	As Left	Max	Dev.
0.00	-0.50		0.50		-0.50		0.50	
25.00	24.50		25.50		24.50		25.50	
50.00	49.50		50.50		49.50		50.50	
75.00	74.50		75.50		74.50		75.50	
100.00	99.50		100.50		99.50		100.50	
75.00	74.50		75.50		74.50		75.50	
50.00	49.50		50.50		49.50		50.50	
25.00	24.50		25.50		24.50		25.50	
0.00	-0.50		0.50		-0.50		0.50	

TEST EQUIPMENT SERIAL OR I.D. # _____

TEST EQUIPMENT CAL. DUE DATE _____

COMMENTS _____

CALIBRATED BY_____ DATE _____

Figure 3.1-1 *Scale Data Sheet/M&TE Tag*

or technique. Since it is impractical and expensive to perform all calibrations in the atmospherically controlled conditions of a laboratory clean room, different levels of standards are used. Primary standards are either traceable directly to an artifact of the basic SI unit quantity, or produced using known physical constants. An artifact is an arbitrarily accepted standard of measure maintained under ideal conditions. Primary standards are maintained as reference standards. Each laboratory routinely verifies the measurement of secondary standards to the primary standard. Secondary standards are clean room laboratory standards maintained as reference standards by the end user. Secondary standards (working standards) are routinely validated to maintain their pedigree. Working standards are compared to the reference standard at predetermined intervals to reestablish and confirm uniformity. Working standards are used to calibrate tertiary or field standards. Field (tertiary) standards are the devices that a technician uses to calibrate permanent plant equipment. In each case, the testing equipment should have a serial number, the date of calibration, the name of the person who performed the last calibration, and a calibration expiration date. Each piece of test equipment should have a permanent data record that identifies the serial number of the standard it was calibrated against, the date of the standard's last calibration, and the standard's calibration expiration date. The record should also include the calibration data from each time the test equipment was re-calibrated. A systematic review of the historical data reveals adverse trends such as increasing deviation over each calibration cycle, and permits accurate determination of the instrument's stability. The calibration cycle time is adjusted or the test equipment is repaired or replaced. Detailed analysis of uncertainties is beyond the scope of this text. A statement of uncertainty is an estimate of all possible error in a measurement, or a stated range of values within which the true value must reside. Uncertainty includes not only the specification of the device, but all possible deviant factors. Eliminating variable components in the calibration process reduces uncertainty. Training personnel, controlling environmental factors, incorporating written calibration procedures, establishing and maintaining a quality assurance program, and reducing the number of times a measurement is transferred all increase certainty and reduce uncertainty. Test equipment found out of tolerance renders all calibrations suspect that were performed with it since the last time it was calibrated. Analysis of the data determines whether or which calibrations require validation. If test equipment is dropped or damaged in any way, stop using it, and report the event. It is then re-certified, and any gross deviations attributed to a single incident. Standards and laboratory calibrations are a specialized field of instrumentation called metrology. It is important for the process technician to understand the validity of traceable field standards.

3.2 Accuracy, Error, and Precision

Calibration is the comparing, adjusting, and recording of a device's response in relation to a known standard. Several terms assist in the description of a particular response. The most common term is probably accuracy. Accuracy is actually used to define the amount or degree to which a device does not replicate the standard, or the maximum amount of error that is expected. Accuracy is stated as percent of the indicated reading, percent of span, or in units of the measured variable. Figure 3.2-1 is a response graph for an accuracy of +/- 10%. This value is stated as units of the measured variable (+/- 2 psig), or as a percent of span or upper range value (+/- 2%). It is often necessary to convert between the methods to determine the appropriate standard to calibrate a particular device. Typical calibration procedures require the use of field standards possessing a minimum of four times the accuracy of the field device under calibration. If a 100 psig pressure gauge requires calibration to accuracy of +/-0.5% of span and a four to one ratio is required, the percent statement is first converted to actual measured variable units. The conversion yields 100 psig × 0.5% = 0.5 psig. Divide by four and determine the measured variable value accuracy for the standard selected. This yields 0.5 psig/4 = 0.125 psig. A 1,000 psig test gauge with an accuracy of +/- 0.25% of span is available. The test

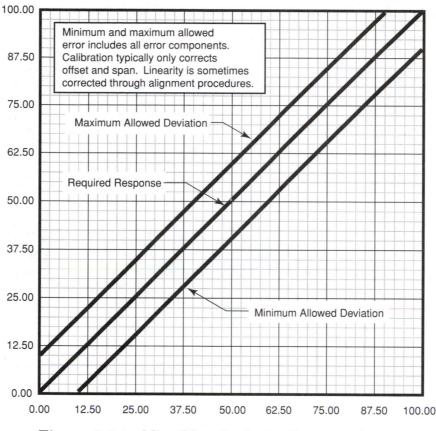

The graph shows axes labeled from 0.00 to 100.00. Text box reads: "Minimum and maximum allowed error includes all error components. Calibration typically only corrects offset and span. Linearity is sometimes corrected through alignment procedures." Labels: "Maximum Allowed Deviation", "Required Response", "Minimum Allowed Deviation".

Figure 3.2-1 *Min./Max. Deviation Response Graphs*

gauge range of 1000 psig is multiplied by the stated accuracy of 0.25% of span resulting in a measured variable value of +/- 2.5 psig. The acceptable error of the test gauge exceeds the acceptable error of the process gauge; obviously, another standard is required. A 1000 psig test gauge with a stated accuracy of 0.01% of span is also available. Converting the percent accuracy statement into the measured variable results in +/- 0.1 psig. The calculation reveals sufficient margin. A 100 psig test gauge with stated accuracy of +/- 0.1% of span is also available. The 100 psig test gauge measured variable is rated at +/- 0.1 psig. Even though the 1000 psig test gauge has a higher degree of stated accuracy in percent of span than the 100 psig test gauge, the 100 psig test gauge not only provides equivalent measured value accuracy but greater resolution as well, and is therefore a better choice. Deadweight testers are often specified in terms of percent of reading. A deadweight tester specified as +/- 0.1 percent reading has a measured variable accuracy of +/- 0.1 psig at 100 psig and a variable accuracy of +/- 1.0 psig at 1000 psig. Resolution is the discernible interval between the least significant marks or identified increments. Do not record indicated values less than one-half the least significant increment. Sensitivity identifies the inherent design capability of a device to detect a change of input value that results in a change in output value. Dead band, Figure 3.2-2, is the amount of change in input required to produce a change in output when reversing the direction of the input. Hysteresis, Figure 3.2-3, is the amount of error produced by approaching the exact same value from increasing and decreasing directions. Unlike deadband, which is detected as a lack of response, hysteresis errors are immediate.

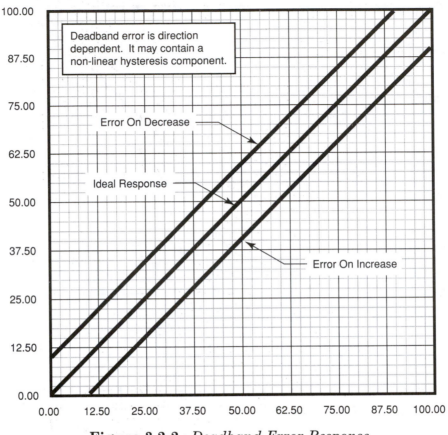

Figure 3.2-2 *Deadband Error Response*

Precision is the degree that a method or device repeats exact duplicate data for each measurement sequence performed. Stability is the ability of a device to maintain accuracy over time. Precision is used to define immediate repetitive measurement, while stability includes drift over a defined period of time. High precision is a desirable trait where repeatability is a higher concern than absolute accuracy. This is often a moot point with modern digital devices, however, design and manufacturing limitations often present a dichotomy with analog devices. Users were forced to choose between accurate but delicate instruments, or rugged devices with good precision and stability but only fair accuracy. Precision of test equipment generally exceeds the skill or method of the person performing the calibration. If testing produces differing results, review and analyze the method or technique prior to making adjustments. Make no adjustments until the source of uncertainty is identified and corrected. A reasonable comparison is a rifle marksman adjusting his rifle sights. His rifle and ammunition, when clamped in a shooting vise, consistently group the shots within a one-inch circle on a one hundred yard target. When the marksman holds the rifle and fires at the same target, the shots scatter randomly all over the target. Any adjustment to the rifle is impossible until the marksman improves his technique. Parallax is an operator-induced error when reading pointer scaled devices. Parallax error is minimized or eliminated by observing the pointer from a perpendicular direction and aligning the pointer with its own reflection in the mirrored scale.

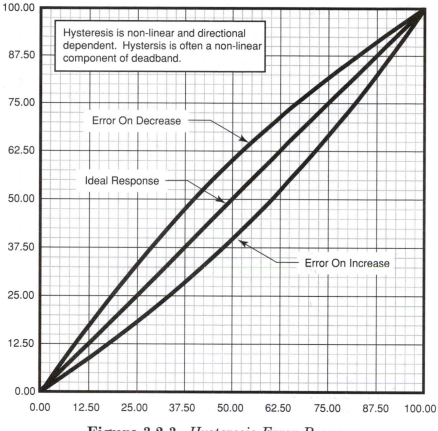

Figure 3.2-3 *Hysteresis Error Response*

3.3 Static and Dynamic Data

There are two primary types of data collected when testing and/or calibrating instrumentation devices and instrumentation systems: static data and dynamic data. Static data is obtained by applying a fixed value and recording data after the device output has stabilized. Static values are not related to time, and generally represent the stated accuracy and performance characteristics of a device. Several data points are selected, typically 0%, 25%, 50%, 75%, 100%, 75%, 50%, 25%, and 0%. Although data is collected at nine distinct points, this is often called a five-point check. If any desired input value is exceeded, return to the previous value and re-approach the data point. The direction of approach is critical to reveal hysteresis. Extremely critical calibrations record data at additional points and some calibrations record data at fewer points. A switch might only have a single significant data point, the last legible value prior to the change of state, see Figure 3.3-1. Good work practice dictates the reset value is recorded as well.

Static calibrations are extremely reliable, and while sometimes applied to the entire loop, they are often performed on one device at a time.

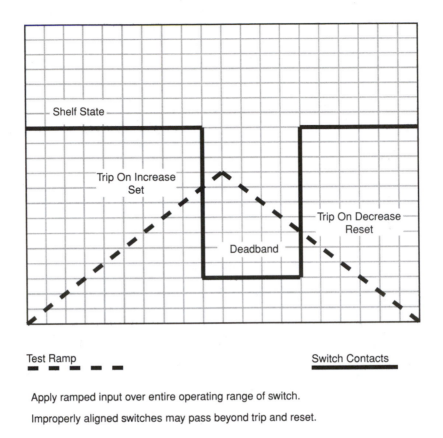

Test Ramp ▬ ▬ ▬ ▬ ▬   Switch Contacts

Apply ramped input over entire operating range of switch.

Improperly aligned switches may pass beyond trip and reset.

Figure 3.3-1 *Switch Response*

Dynamic testing is a more reliable method to disclose the operational characteristics of a device or system under actual process conditions. Dynamic calibrations measure time weighted response characteristics of a loop or device, and require the use of recorders and step or ramp generators. Measured inputs are applied as either a series of steps or as a steadily increasing/decreasing ramp. The outputs from the device under calibration are monitored and recorded continuously throughout the test. The data points are analyzed at the same values as the static calibration, but additional information is available, see Figures 3.3-2, 3.3-3, and 3.3-4. In addition to traditional errors, dynamic methods also disclose time response characteristics such as ramp time, derivative time constant, decay time, integral time constant, overshoot, and oscillation. Time is a critical parameter of dynamic calibrations. Different test times produce different results from the same loop or component. Analog calibration test times are selected in relation to loop response times. Testing times identify the test signal response slope and are identified in terms of slope, volts per minute or milliamps per minute, or as total ramp time required for the specified input to change between the minimum and maximum test values, "4 – 20 milliamps in two minutes." Safety actuation test time is actual elapsed time from the initiating event to the final response or ultimate condition. The same loop or device develops different response characteristics under different testing times.

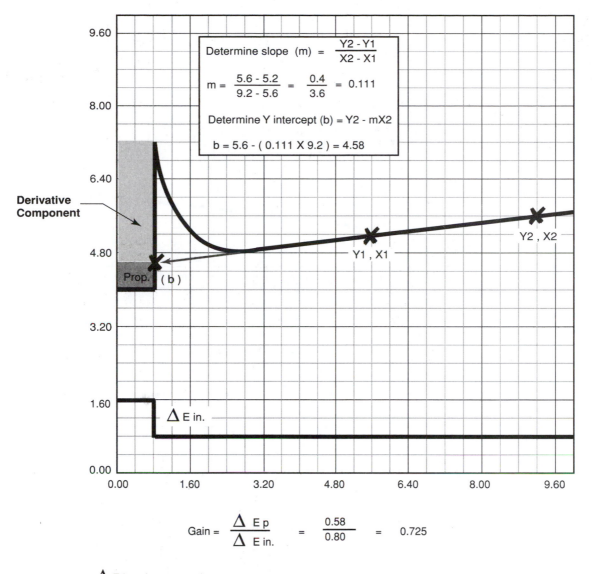

Determine slope (m) = $\dfrac{Y2 - Y1}{X2 - X1}$

m = $\dfrac{5.6 - 5.2}{9.2 - 5.6}$ = $\dfrac{0.4}{3.6}$ = 0.111

Determine Y intercept (b) = Y2 - mX2

b = 5.6 - (0.111 X 9.2) = 4.58

Derivative Component

Y2 , X2

Y1 , X1

Prop. (b)

Δ E in.

Gain = $\dfrac{\Delta \, E \, p}{\Delta \, E \, in.}$ = $\dfrac{0.58}{0.80}$ = 0.725

Δ E in. = input step change = input initial - input final = 1.60 - 0.80 = 0.80

Δ Ep = Proportional component = b - V output initial = 4.58 - 4.00 = 0.58

Figure 3.3-2 *Dynamic Loop Response (PID) Proportional Component*

Calibrations are performed independently on each individual device in the loop (device calibration), or on the entire collective loop simultaneously (loop calibration). Monitoring the entire loop simultaneously ensures that all interconnecting wiring is evaluated and that additive error is included in the data. Power supply loading and noise are also more pronounced and evident under loop calibrations. Additive error is a phenomena induced by transmitting the same signal through several connected devices. Each device in the loop adds some error to the signal, and as the signal continues along the transmission path it propagates each accumulated error with it, see Figure 3.3-5. Random generated errors in many cases will statistically eliminate one another, providing the

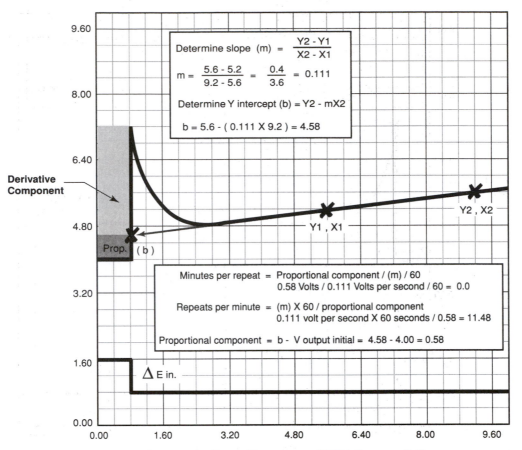

Figure 3.3-3 *Dynamic Loop Response (PID) Integral Component*

amount of positive error is roughly equivalent to the amount of negative error. Engineering analysis determines the criticality of the control system and specifies instrument accuracies to minimize this effect.

Regardless of the calibration technique selected, precise values are only obtained by precise techniques. It is critical to perform calibrations by the most repetitive means possible. Several factors influence the validity of recorded values. As previously mentioned, all testing equipment should be traceable to a known standard, in good working condition, and used within its prescribed range of accuracy. Other factors considered are temperature, vibration, and humidity. Obviously field "in-situ" calibrations often-predicate measurement under such adverse conditions. It is important to review the manufacturer's stated environmental limits of test equipment operability and to use the equipment accordingly. Place the testing equipment in a desirable environment and install remote connecting cables or tubing if required. Several human factors also affect repeatability. Proper lighting is necessary for legibility. Always place testing equipment in a stable and easy to read position. Try to orient yourself in a comfortable position.

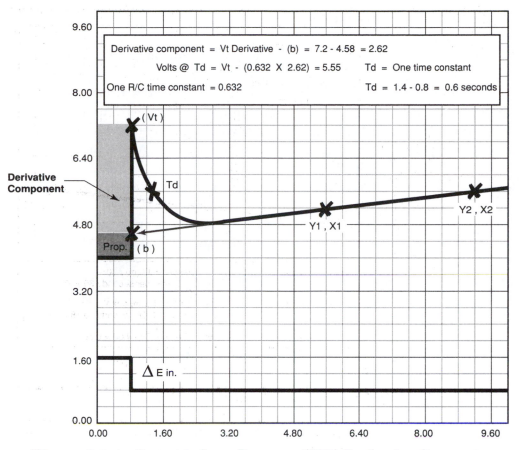

Figure 3.3-4 *Dynamic Loop Response (PID) Derivative Component*

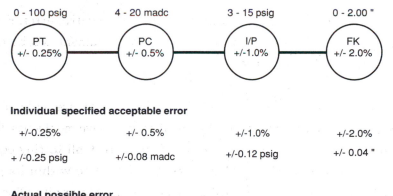

Individual specified acceptable error

+/-0.25%	+/- 0.5%	+/-1.0%	+/-2.0%
+ /-0.25 psig	+/-0.08 madc	+/-0.12 psig	+/- 0.04 "

Actual possible error

+/-0.25%	+/-0.75%	+/-1.75%	+/- 3.75%
+ /-0.25 psig	+/-0.12 madc	+/-0.21 psig	+/-0.075"

Although each individual device is within specified tolerance,

Additive error cumulatively induces total out of tolerance response.

Figure 3.3-5 *Additive Error*

73

3.4 Adjustment Techniques

Several types of adjustments are generally available. The most basic adjustments are offset and multiplication. Offset adjustments do not change the slope or response characteristics of the device, see Figure 3.4-1. Offset, also known as zero or bias, shifts the entire response up or down. Most instruments have an external adjustment screw or port labeled as "zero" that is used to tune this parameter. It is typically adjusted at the lower range input value and fixes or establishes the lower range value output response, or the starting point of device response. Adjust the potentiometer which connects the amplifier reference input to circuit ground to correct the offset of electronic devices. Adjust the scale or pointer to correct offset error in linkage type devices. Adjust the balance beam offset spring or the flapper nozzle gap to correct offset in pneumatic beam devices. Some bipolar devices (devices that output a positive and negative response about a center or median value) establish the offset or zero value at the midpoint of total span.

Multiplication adjustments fix or determine the characteristic response slope of the output in relation to a change of input value. Multiplication adjustments are often labeled span, range, or gain.

> *Even though the terms span, range, and gain are all equally acceptable to describe a multiplication adjustment, it should be noted that span and range are actually the inverse of gain and multiplication. Decreasing gain increases span, and increasing span decreases gain.*

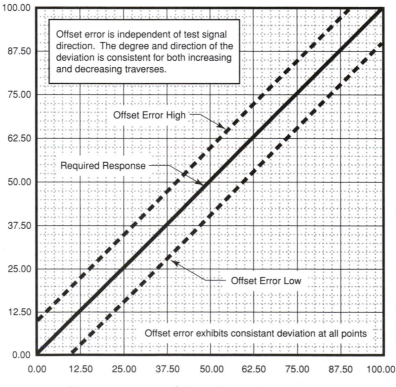

Figure 3.4-1 *Offset Error Response*

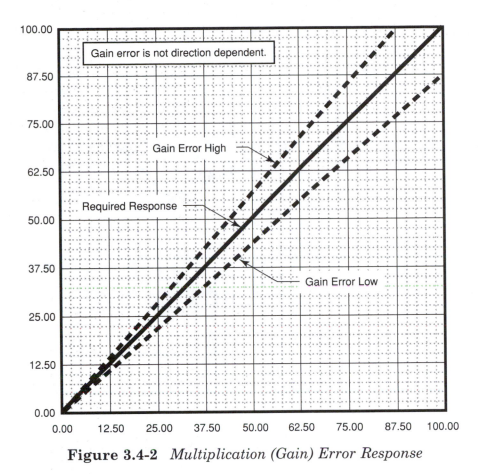

Figure 3.4-2 *Multiplication (Gain) Error Response*

Multiplication adjustments are typically performed at the upper range value, and inferentially fix the upper value as a result of input value multiplication, see Figure 3.4-2.

Increase or decrease the value of the feedback potentiometer circuit to effect electronic multiplication. The feedback potentiometer is located in the circuit that connects the amplifier output back to the measuring input. Higher resistance values result in greater multiplication (gain). To adjust multiplication on a linkage device, shift the distance between the drive arm connection at the driven arm and the pivot point of the driven arm. Moving the drive arm closer to the pivot point increases the multiplication and produces a greater output response in relation to input response. Pneumatic balance beams have an adjustable range nut that determines the mechanical advantage of the beam. Shift the fulcrum closer to the flapper/nozzle to reduce multiplication. In most cases, the proper adjustment is the amount of correction required to eliminate one-half of the error at the upper range value. Additional offset adjustment is usually required to reestablish zero indication. See Figures 3.4-3, 3.4-4, and 3.4-5.

A linear response generates an equivalent change in output for each fixed change in input. The slope of the response graph is a straight line. Adjusting offset shifts the entire line up or down; adjusting gain changes the slope of the line. Non-linearity is evident as a mid-span deviation with the offset and multiplication values corrected. Electronic devices are selected to operate within their characteristic

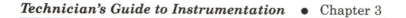

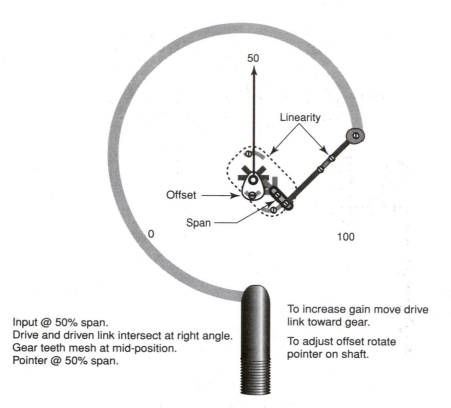

Figure 3.4-3 *Adjustable Linkage Assembly*

region of linearity and seldom require adjustment. Mechanical linkages, however, only exhibit linear characteristic if aligned properly. Most mechanical assemblies provide some type of linearity adjustment. Exact alignment procedures are found in the respective manufacturer's literature.

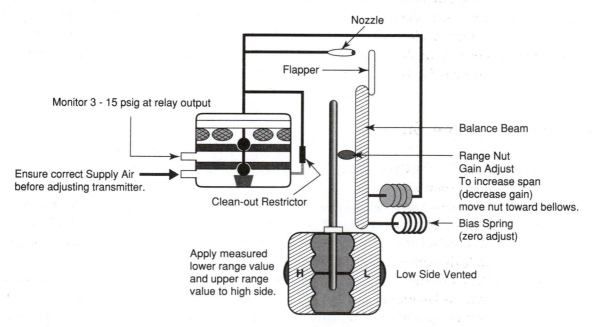

Figure 3.4-4 *Balance Beam Adjustment*

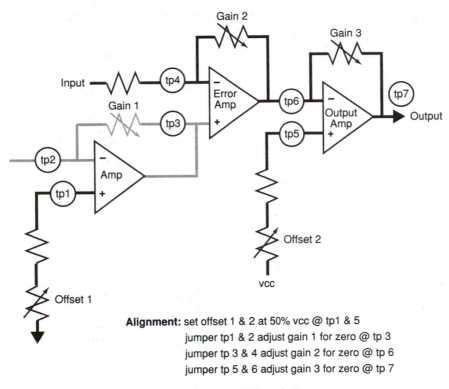

Figure 3.4-5 *Amplifier Adjustment*

The alignment procedures are typically performed with a mid-span input value applied to the device. Offset and multiplication adjustments are placed in the middle of their adjustment range. Any gear or cam assemblies are adjusted to their middle position, and the intersect angle of connecting linkages is adjusted to ninety degrees. Fine tuning of mechanical link linearity error is accomplished by adjusting the intersect angle of the connecting link as required to move the indication five times the amount of the error in the direction of the error. Additional offset and multiplication adjustment is often required to reestablish the zero point indication.

■ Chapter Three Summary

- The standards custodian for the United States is the National Institute of Standards and Technology, formerly known as the National Bureau of Standards.
- Calibration is the process of comparing the outputs of a device to a known traceable standard, and/or adjusting the outputs of a device to replicate the standard within a specified tolerance.
- A calibration cycle consists of multiple point comparisons to a standard over the entire range of the device from both increasing and decreasing directions.
- Trace ability is the documentation process that validates the lineage of a measured value back to a national standards laboratory.
- A calibration report is a graph or table that contains the comparison data between a standard and the device under calibration.

- The calibration report should include the serial number, calibration date, calibration expiration date of the standard used, the desired value of all data points monitored with minimum and maximum acceptable values, and data blanks to record as found and as left readings.
- Primary standards are those standards maintained by the governing body or those that are reproducible using known physical constants.
- Secondary standards are generally clean room laboratory standards owned and controlled by the end user.
- Field (tertiary) standards are the devices that a technician uses to calibrate permanent plant equipment.
- Each piece of test equipment should have a permanent data record that identifies the serial number of the standard it was calibrated against, the date of the standard's last calibration, and the standard's calibration expiration date. The record should also include the calibration data from each time the test equipment was re-calibrated.
- Calibration is the comparing, adjusting, and recording of a device's response in relation to a known standard.
- Accuracy is used to define the amount or degree to which a device does not replicate the standard, or the maximum amount of error that is expected.
- Resolution is the discernible interval between the least significant marks or identified increments.
- Sensitivity identifies the inherent design capability of a device to detect a change of input value that results in a change in output value.
- Deadband is the amount of change in input required to produce a change in output when reversing the direction of the input.
- Hysteresis is the amount of error produced by approaching the exact same value from increasing and decreasing directions. Unlike deadband, which is detected as a lack of response, hysteresis errors are immediate.
- Precision is the degree that a method or device repeats exact duplicate data for each measurement sequence performed.
- Static data is obtained by applying a fixed value and recording data after the device output has stabilized.
- Dynamic testing is a more reliable method to disclose the actual operation of a device or system under actual process conditions. In addition to traditional errors, dynamic methods also disclose time response characteristics such as ramp time, decay time, overshoot, and oscillation.
- Additive error is a phenomena induced by transmitting the same signal through several connecting devices.
- It is important to review the manufacturer's stated environmental limits of test equipment operability and to use the equipment accordingly.
- Offset adjustments do not change the slope or response characteristics of the device.
- Multiplication adjustments fix or determine the characteristic response slope of the output in relation to a change of input value.
- A linear response generates an equivalent change in output for each fixed change in input. The slope of the response graph is a straight line. Nonlinearity is evident as a midspan deviation with the offset and multiplication values corrected.

Chapter Three Review Questions:

1. Identify the standards custodian for the United States of America.

2. Define the calibration process.

3. Define a calibration cycle.

4. State the advantage of traceability.

5. List the critical information provided by a calibration report.

6. Differentiate between artifact standards, intrinsic standards, secondary standards, and tertiary standards.

7. List the minimum information affixed to each item of test equipment.

8. List the permanent data record information maintained for each piece of test equipment.

9. State the definition of resolution.

10. State the definition of sensitivity.

11. State the definition of deadband.

12. Explain the difference between deadband and hysteresis.

13. Explain the difference between precision and accuracy.

14. Explain the difference between static data and dynamic data.

15. Explain additive error.

16. Analyze and interpret a calibration response graph to identify offset error, multiplication error, and linearity error.

17. Analyze and interpret a calibration response graph to determine adjustments required to eliminate offset error, multiplication error, and linearity error.

18. Analyze and interpret a schematic diagram to determine adjustment locations required to eliminate offset error and multiplication error.

19. Analyze and interpret a mechanical lever mechanism to determine adjustment locations required to eliminate offset error, multiplication error, and linearity error.

20. Analyze and interpret a force balance beam assembly to determine adjustment locations required to eliminate offset error and multiplication error.

Chapter 4

Signal Characteristics and Test Equipment

OBJECTIVES

Upon completion of Chapter Four the student will be able to:

- List the common testing devices used to detect and identify electrical and electronic signal characteristics.
- List the common electrical signal characteristics identified and manipulated during troubleshooting or calibration activities.
- Compare and contrast the differences between using a digital multimeter in the volt, ohms, and ampere position.
- State the precautions required when connecting a digital multimeter into a circuit to monitor amperes or ohms.
- Explain the "live, dead, live" safety practice.
- Determine the appropriate testing equipment required to measure 4 – 20 mADC as 1 – 5 VDC.
- Determine the appropriate testing equipment to monitor for an intermittent signal error.
- State the appropriate grounding precautions when installing a recorder.
- State several precautions required when using a megohmmeter.
- State several uses and advantages of clamp-on ammeters.
- State several signal characteristics detected with an oscilloscope including frequency, period, pulse width, voltage ACp-p, volts AC RMS, volts DC offset, rise time, decay time, and noise.
- Compare and contrast floating ground systems and conventional earth ground systems.

- *State several precautions required when working with or handling static sensitive components.*
- *List several testing devices used to detect, simulate, or measure pressure.*
- *State several safety considerations and precautions when calibrating a pressure device in the field.*
- *Explain the proper selection, use, and operation of a test gauge.*
- *Explain the proper use and operation of digital pressure testers.*
- *Explain the benefits and limitations of using a wet-leg calibrator.*
- *State several safety considerations and precautions when calibrating a level device in the field.*
- *Explain several methods used to perform temperature calibrations.*
- *State the purpose of an ice-point reference cell.*
- *Determine the thermocouple millivolt equivalent for various temperature values.*
- *Determine the RTD ohmic equivalent for several temperature values.*
- *Identify each thermocouple type by the thermocouple color code.*
- *State several safety considerations and precautions used when performing temperature calibrations.*
- *State several precautions to prevent introduction of incompatible fluids into the process.*

4.0 Introduction

A myriad of testing equipment is available for the technician to use. This text covers a representative range of common test devices. The value under determination or investigation determines the test equipment selected. Monitoring a live operational signal and comparing the result to the expected value is a common troubleshooting technique. When performing calibrations, the technician must generate the test signal value. This chapter includes common testing equipment and several common test apparatus used to simulate the measured variable.

4.1 Sinusoidal Waveform Analysis

The oscilloscope monitors voltage in respect to time or compares two voltages.Oscilloscopes display a waveform of the monitored signal. Required value units are extracted from the waveform display and multiplied by the oscilloscope dial settings. Oscilloscopes typically come with sensing probes. Prior to using the probe, attach it to the probe adjust tab on the oscilloscope dial and tune the probe to obtain a square wave pattern. The probes are selectable for 1X or 10X. Probes in the 10X configuration attenuate (reduce) measurement magnitude by a factor of ten.

All common definitions of electrical signal characteristics are evident in the sinusoidal waveform, Figure 4.1-1. The horizontal distance from peak to peak is one time period.

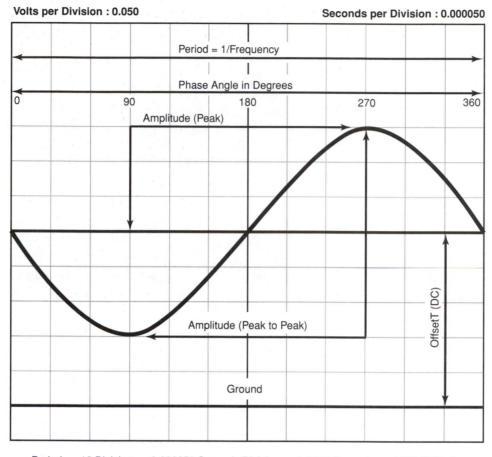

Volts per Division : 0.050　　　　　　　　　　　**Seconds per Division : 0.000050**

Period = 12 Division × 0.000050 Seconds/Division = 0.0006 Seconds = 1666.66 Hertz

Volts (p-p) =　6 Division　× 0.050 Volts/Division = 0.3 $V_{p\text{-}p}$

Volts (Peak) =　3 Division　× 0.050 Volts/Division = 0.15 V_p

Volts RMS =　Volts peak × 0.707 = 0.106　　　Volts p-p = Volts RMS × 1.414 = 0.15
Offset (DC) =　5 Division　× 0.050 = 0.25

Figure 4.1-1　*Sinusoidal Wave Form*

Frequency is the reciprocal of time period. The vertical distance from positive peak to negative peak is amplitude peak to peak or in the case of electrical measurement voltage peak to peak (Vp-p). Peak to peak voltage divided by two is peak voltage (V_p). Peak voltage multiplied by 0.707 is voltage root mean sum, (RMS); conversely volts RMS multiplied by 1.414 is equal to peak voltage. Voltage statements without subscript identifiers are conventionally assumed as RMS. If two sine waves are displayed simultaneously the distance between the adjacent peaks is the phase differential. The distance between the centerline of the sine wave and earth ground is offset. Noise is an unwanted interference signal that carries no usable data. Very high frequency noise is often apparent as a diminutive sinusoidal signal riding the actual signal. Very low frequency noise is not readily apparent, as it tends to shift the offset of the entire signal waveform at the noise frequency. Additional signal information available from an oscilloscope display includes rise time, decay time, and noise. Although the popular definitions of period and phase are referenced from amplitude peak to amplitude peak, more accurate measurement is calculated from the point where the waveform intersects the zero reference line.

4.2 Electronic Testing Devices

Electronic test equipment includes digital multimeters, signal generators, precision resistors, recorders, megohmmeter, clamp-on ammeters, loop calibrators, and oscilloscopes. They all detect and measure specific characteristics of electricity, such as frequency, volts, amperes, and resistance. Digital multimeters (DMMs) are the most versatile. They monitor volts AC (rms), volts DC, ohms, amperes, and frequency. Since multimeters are capable of measuring multiple

Figure 4.2-1 *Digital Multimeter**

parameters, it is critical to configure the multimeter properly before connecting to the circuit. Multimeters placed in the voltage function exhibit very high impedance and can usually measure voltage without disrupting the circuit. Figure 4.2-1 demonstrates a DMM used to measure volts. Figure 4.2-2 is a frequency counter.

Prior to working on an electrical component, determine if it is energized using the "live-dead-live" technique. First determine the operability of the multimeter by measuring a known energized source. Next, verify the state of the component under investigation. Last re-verify the multimeter by measuring a known energized source.

Multimeters in the ampere or ohm function introduce a direct short into the circuit and can cause serious component damage if improperly installed.

It is good practice to verify test leads prior to use. Good leads should measure almost zero ohms when shorted together, and infinite ohms when separated. Coaxial cable and twisted pair leads offer fair noise resistance. Use no more lead length than necessary. Route leads

*Photo Courtesy of Fluke

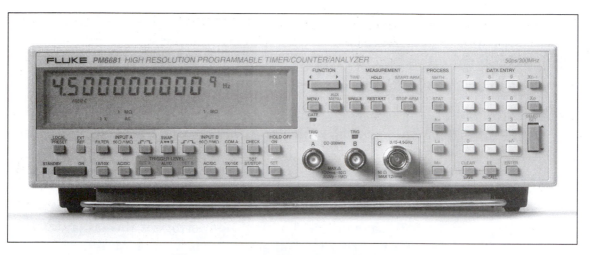

Figure 4.2-2 *Frequency Counter**

away from high-energy field and devices. Signal generators are often used as a source of the known test signal. Many signal generators require AC line voltage to operate. Use great care to determine the required isolation before connecting into the circuit. Precision resistors and decades are most commonly used to convert a current signal into a voltage, or to simulate an RTD. They are placed in series with the current loop and the voltage drop across them is monitored with a voltmeter. Recorders, megohmmeters, and clamp-on ammeters are useful troubleshooting devices. Recorders are used to collect dynamic data or to monitor a point for intermittent deviations. Most recorders offer program capabilities to scale the graph, select chart speed, and set trigger threshold voltage. Megohmmeters measure insulation breakdown voltage and are commonly used to detect faulty cabling.

Figure 4.2-3 *Clamp on Meter**

*Photos Courtesy of Fluke

High voltage is stored on tested cables. Ensure the megohmmeter is properly drained prior to disconnecting from a recently tested cable.

Megohmmeters output sufficient voltage to destroy electronic circuitry. Do not attempt to megohm cables unless they are disconnected from all circuits.

Clamp-on ammeters, Figure 4.2-3, are used for non-intrusive current readings and are useful to detect grounding faults.

Loop Calibrators combine both the input and output monitoring for the calibration, Figure 4.2-4 and 4.2-5. They contain an onboard signal generator and the required transducers to measure both the input and output values. Documenting calibrators accept user programmed test parameters and automatically apply and record the output and input values. The calibration data is easily downloaded to a laptop computer that contains the appropriate calibration record software, and or printed to a floppy disk or CD. Many modern instruments are HART compatible. HART is the acronym for highway addressable remote transducer. This is an industry standard that regulates communication protocol and allows information exchange among devices connected to a common 4 – 20 mADC current loop signal. As opposed to conventional calibrations, which monitor the signal directly at the

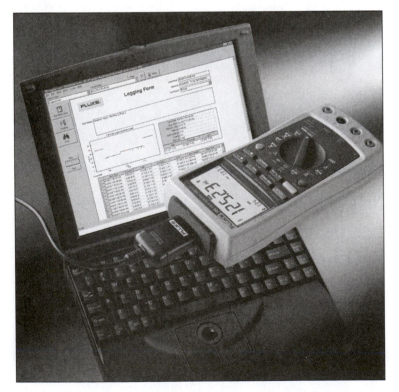

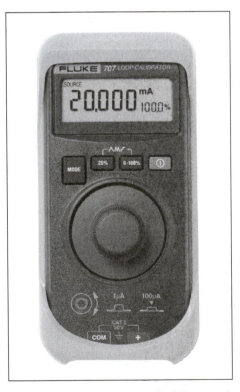

Figure 4.2-4 *Documenting Calibrator**   **Figure 4.2-5** *Loop Calibrator**

*Photos Courtesy of Fluke

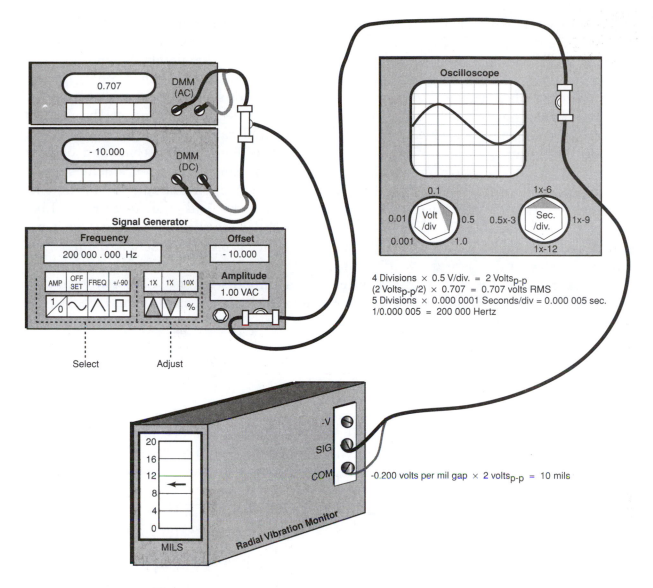

Figure 4.2-6 *Vibration Monitor Calibration Set-up*

device, HART compatible devices are remotely monitored at any point along the current loop. In addition to calibration adjustment, several other parameters are monitored and adjusted. Additional specifics are provided in a later chapter. Figure 4.2-6 demonstrates a practical calibration set-up used to calibrate vibration-monitoring equipment.

4.3 Isolation, Static, and Grounds

All electrical and electronic testing and measurement activity requires particular awareness of isolations, grounds, and static discharge potentials. Many control systems operate with a floating or isolated ground system. This means that while all components of the particular system are electrically tied to a common reference, that reference is not tied to plant common or earth ground. The electrical potential of a floating ground, common, is often substantially different from an adjacent earth ground. Completely isolate any test equipment used on such systems from plant common ground or earth ground. Exercise particular care when attempting to simultaneously monitor several separate systems with a single device, such as a recorder. The recorder may electrically connect the separate floating ground potentials of each system together and cause serious damage. Battery powered test equipment is generally not a problem as long as the technician does not connect a test lead to earth ground. Instrument signal cables are often shielded with a grounding drain strand. The preferred installation method is to terminate the drain shield at the control cabinet back plane and thoroughly isolate the field end. Never ground both ends of an instrument cable shield. Metallic conduit and raceways, however, are considered guards and are earth grounded at both ends. Good wiring practices include using twisted pair signal wires, using a drain or shield landed at the control cabinet back plane, reducing or eliminating circuit commons and daisy chains, adding fuses to isolate potential individual faults, separating high energy power cables from instrument signal cables, and installing guards for radio frequency (Rf) sensitive components. See Figure 4.3-1.

> **NOTE** *Common isolation transformers that use a three-prong outlet do not offer complete ground isolation.*

Many test devices are wired with an integral case ground. A case grounded device placed on a metal grate can provide a path to earth ground. Electrostatic discharge can also cause irreparable damage to electronic circuits. Lightening is an exaggerated example of electrostatic discharge. Each of us routinely charge and discharge static potential on a

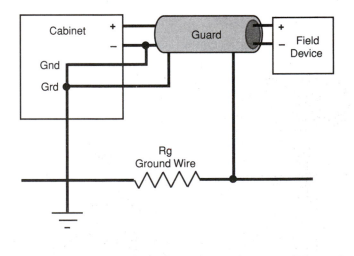

Guards: pipe and metallic conduit should be grounded at both ends.

Signal common and grounds should terminate at cabinet back plane.

Due to Rg some voltage difference is always present at remote grounds.

Figure 4.3-1 *Grounds and Guards*

continual basis. Many sensitive areas are humidity controlled to minimize the amount of static charge created. Always use anti-static mats and wrist straps when working with static sensitive components. Replacement circuits are shipped from the factory with anti-static packaging material. It is a good practice to store the circuits in the original protective packaging until ready for use. Do not remove the device from the packaging without anti-static safeguards in place. Connect the wrist strap to the anti-static mat when unpacking components or configuring circuit cards. Connect the wrist strap to the card frame when removing and replacing circuit cards. Always touch the common frame before handling individual components. Individuals should grasp hands prior to transferring static items from one to another.

4.4 Pressure Testing Devices

The type of signal injected or detected determines the device selected. Calibration requires the comparison of a known good value to the inspected value. Multiple devices are often required to generate and measure the reference value. One device might inject the desired input value, while another monitors the input value, and a third device monitors the output value. Figure 4.4-1 depicts a test gauge calibrated with a dead weight tester. Figure 4.4-2 depicts a typical wet-leg calibration set-up for a flow transmitter.

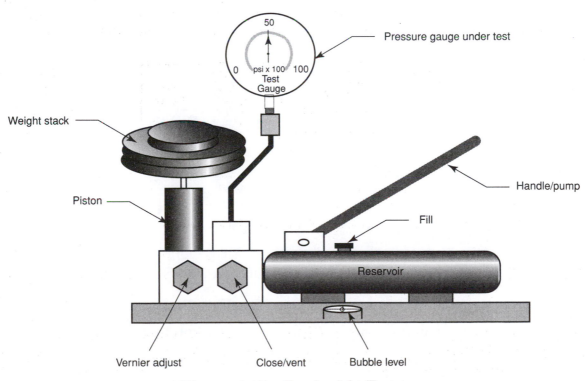

Figure 4.4-1 *Deadweight Tester*

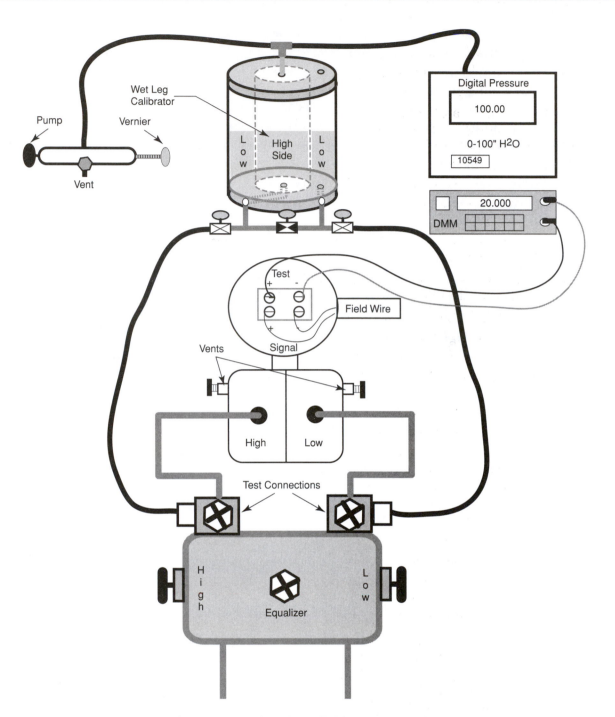

Figure 4.4-2 *Wet Leg Calibrator Test Set-up*

Pressure testing equipment includes deadweight testers, test gauges, manometers, digital pressure detectors, hand pumps, regulators, and wet-leg calibrators. Deadweight testers are hydraulic pumps that parallel fluid pressure to the output fitting and a precision-machined piston assembly. Accuracy is extremely high. Values of +/- 0.1% of reading are common. Increased accuracies are available. The piston assembly has a precision weight set comprised of various weight values. The desired pressure value is determined by stacking the selected weights on the piston. The device under test is connected to the output fitting and the tester handle is pumped until the piston lifts the weight stack. The pressure required to suspend the weight stack is simultaneously applied to the output port. Friction forces are minimized as long as the weighted piston is spinning, typically ten to thirty revolutions per minute. Two adjustment knobs are provided. The drain valve knob is used to vent large increments of pressure. The other knob is an adjustable vernier that allows fine-tuning of the pressure. The piston plunger is a much smaller diameter than the piston and weights; therefore, the weights are stamped with representative values, not their true actual weight. Although low pressure, inches of water, deadweight testers are available, commonly used testers measure up to ten thousand pounds per square inch. Since the fluid in the tester is applied to the device tested, ensure compatibility of the testing fluid to the process media, and/or flush the device prior to installation. Each deadweight tester is supplied with its own serialized weight set. Maintain the weights clean and free from nicks, dents, and scratches. Handle the weights with gloves to protect them from body oils and salt. Do not interchange weight sets. The piston and cylinder dimensions establish the inherent accuracy of a dead weighttester. At maximum pressure, the piston should fall at less than 0.1 inches per minute. Excessive fall rates are indicative of piston or cylinder wear.

Test gauges are similar to process gauges, the primary difference being a more precise manufacturing process and a mirrored anti-parallax dial face. Test gauges are very commonly used in the field. Determine the correct resolution and accuracy prior to use, and exercise caution not to cross contaminate the process with residual fluids left in the gauge from previous use. Most test gauges are sensitive to orientation and require zero adjustment if repositioned.

Manometers are used to measure low-pressure values. They are more commonly used as a bench-testing device. The particulars of their use are addressed in another chapter. Digital pressure testers are commonly used to measure low-pressure values in the field. They are extremely accurate and more convenient to use than manometers. Since they digitally display the pressure values, they eliminate parallax and resolution errors and offer higher precision than manometers. Allow sufficient warm-up prior to use, ensure the reference vent is open, do not use fluid as a calibration media, ensure the internal transducer is kept clean and dry, and do not over range the calibrator.

Hand pumps are often used to develop the test signal input pressure. Use caution not to introduce contaminates into the system from the hand pump. Hand pumps and test gauges are often designated and labeled for specific systems. Designations include contaminated demineralized water, noncontaminated demineralized water, lube oil, hydraulic fluid, and fuel oil. Pressure regulators are sometimes used as a pressure source for calibrations. This is an acceptable practice for constant bleed applications, but it is important to remember that regulators automatically compensate for leakage and as such will mask most minor leaks. When adjusting permanently installed plant regulators provide some weepage. Regulators adjusted against a dead head may change significantly when returned to service. The preferred technique is to adjust regulators with full rated, or expected, flow volume demand, and complete the final adjustment in the increasing pressure direction. Wet-leg calibrators are often used to calibrate boiler feed and level transmitters in the field. Wet-leg calibrators are a dual chamber fluid reservoir with separate high side and low side pressure inlet and outlet ports. The calibrator allows wet calibrations while keeping the hand pump and test gauge isolated from the process fluid. Proper use requires cycling or flushing the reservoir fluid through the test tubing and transmitter until all air is removed. Since the high side and low side are both filled, the tubing length and elevation are self-canceling.

 Do not introduce incompatible calibration fluids into the process.

4.5 Level Testing Devices

The preferred method of level calibration is to vary the level of the monitored process by a known amount and observe the expected device output. It is often impractical to vary the actual process level to facilitate a calibration. The most common technique is to isolate the device from the process and inject a process compatible fluid as required to simulate a change in process level.

CAUTION **Standpipes may contain hazardous, high pressure, or high temperature process media. Exercise extreme caution when isolating, venting, and draining standpipes.**

Attach a tee fitting at the standpipe drain connection. Apply a regulated pressurized volume of compatible fluid to the tee at the standpipe drain. Route a section of clear tubing from the drain tee up to the standpipe vent. The level of fluid in the standpipe is read from the clear tubing. It is important to use compatible fluids with similar density. Ensure the correct point of reference is used to determine calibrations. Figure 4.5-1 depicts this set-up with single valve isolation for a low-energy system. High energy or hazardous systems require double valve isolation. The most common reference is a metal tag affixed to the centerline of the vessel. All data points are referenced as plus or minus, above or below the tag. Some levels are referenced to the grating elevation or the centerline of the level pot itself. The correct values are available on the original design drawing. If it is not practical to vary the fluid level with the device in place, remove the device from the field installation and place it in a calibration well. Ensure all system piping and flanges are properly covered to prevent the intrusion of foreign material. A calibration well is a shop mounted standpipe with vent and drain connections already attached. Dry shop calibrations are performed with a calibrated weight set substituted for the float or displacer assembly. Determine or correct the weight set to the equivalent process fluid density.

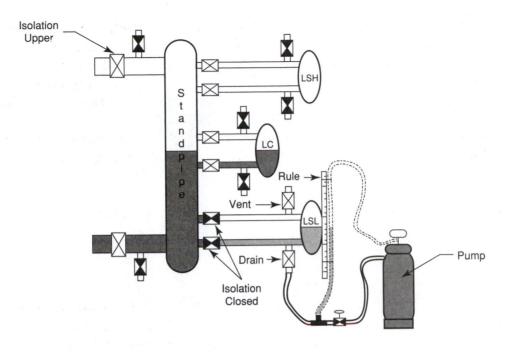

Double valve isolation is required for all
high energy and hazardous systems.

Figure 4.5-1 *Level Switch Calibration Set-up*

4.6 Temperature Testing Devices

Temperature calibrations are often performed with dry block temperature calibrators. These calibrators have a precision control circuit that maintains an accurate well temperature over a wide range of readings, and a set of well adapters to accommodate common probe diameters. Most temperature calibrators offer heat and cool capability. Heated fluidized baths are also used. The heat transfer media in the bath is usually sand, silicone fluid, or water depending on the temperature desired. A laboratory thermometer is used as a temperature standard. It is important to continually agitate a fluidized temperature bath to obtain consistent values. Submerge the probe to an appropriate depth, but do not allow the probe tip to contact the wall or bottom of the bath. Temperature takes time. All temperature elements exhibit some time delay. Allow sufficient propagation and stabilization time prior to obtaining a reading. A minimum of fifteen minutes is often required to allow complete stability. Temperature switches are most reliably adjusted by determining with certainty the temperature at which they will not change state. Incrementally approach the desired set-point and allow sufficient transfer time. The last stable value observed prior to the change of state is the correct set-point. Figure 4.6-1 demonstrates a typical indicator and switch calibration set-up.

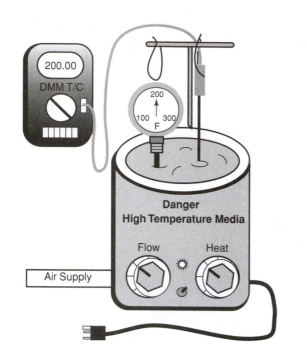
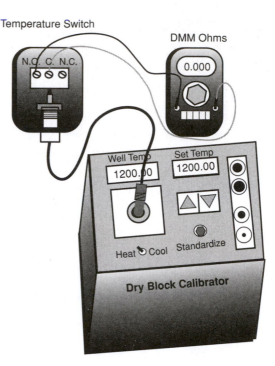

Fluidized bath temperature indicator calibration set-up

Dry block temperature switch calibration set-up

Figure 4.6-1 *Typical Temperature Calibration on Methods*

CAUTION **Use appropriate safety equipment. It is a good practice to erect an exclusion barrier around the temperature bath to prevent inadvertent or unauthorized exposure. High temperature fluids can react violently to the introduction of other fluids. Unsuitable fluids may produce harmful, explosive, or flammable vapors at elevated temperatures.**

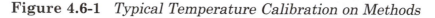

NOTE *Always determine the applicability of the fluidized bath medium for the temperature range expected. Water and other fluids will not exceed their freezing or boiling point temperatures at atmospheric pressures regardless of the energy applied.*

4.7 Electronic Temperature Devices

Electronic temperature detecting devices detect the signal from a field mounted thermocouple or RTD and convert it into a usable transmission signal or display. Thermocouples and RTDs are extremely stable devices and generally exhibit only hard failures. A single point reading of the existing installed thermocouple or RTD can prove operability. Once the field portion of the device is determined functional, the electronic detection circuits are calibrated by simulating the thermocouple or RTD input with a precision voltage or resistance device. RTD inputs are easily simulated with a precision resistor or decade box. Thermocouple inputs are dependent on reference junction

Degrees in Celsius   **Reference Junction at 0 degrees C**

Deg. C	0	1	2	3	4	5	6	7	8	9	10
0	0.000	0.059	0.118	0.176	0.235	0.295	0.354	0.413	0.472	0.532	0.591
10	0.591	0.651	0.711	0.770	0.830	0.890	0.950	1.011	1.071	1.131	1.192
20	1.192	1.252	1.313	1.373	1.434	1.495	1.556	1.617	1.678	1.739	1.801
30	1.801	1.862	1.924	1.985	2.047	2.109	2.171	2.233	2.295	2.357	2.419
40	2.419	2.482	2.544	2.607	2.669	2.732	2.795	2.858	2.921	2.984	3.047
50	3.047	3.110	3.173	3.237	3.300	3.364	3.428	3.491	3.555	3.619	3.683
60	3.683	3.748	3.812	3.876	3.941	4.005	4.070	4.134	4.199	4.264	4.329
70	4.329	4.394	4.459	4.524	4.590	4.655	4.720	4.786	4.852	4.917	4.983
80	4.983	5.049	5.115	5.181	5.247	5.314	5.380	5.446	5.513	5.579	5.646
90	5.646	5.713	5.780	5.846	5.913	5.981	6.048	6.115	6.182	6.250	6.317
100	6.317	6.385	6.452	6.520	6.588	6.656	6.724	6.792	6.860	6.928	6.996

Thermoelectric Voltage in Millivolts

Thermocouples monitored at ice point convert directly from table

Thermocouples monitored at ambient add ambient millivolt equivalent to millivolt observed then find millivolts on table

Thermocouple at ambient reference junction reads 3.364 millivolts

Ambient is measured with thermometer at 22° C

22° C on table is equivalent to 1.313 millivolts

Measuring junction senses 3.364 + 1.313 = 4.677

4.677 millivolts on table is equivalent to 75.5° C

Figure 4.7-1 *Temperature-EMF Table for Type-E Thermocouple*

temperature. An ice-point device is required to provide a 32 degree Fahrenheit reference junction. Thermocouple tables, Figure 4.7-1, relate millivolts EMF to temperature across the entire usable range of each type thermocouple. Thermocouple types are identified by the color of their insulating jacket. Similar tables are used to determine the amount of ohms resistance expected for each temperature value of an RTD, Figure 4.7-2. Figure 4.7-3 demonstrates a typical thermocouple and RTD test set-up.

deg C	Ohms	deg C	Ohms	deg C	Ohms	deg C	Ohms
-25	90.15	5	101.95	35	113.61	65	125.16
-20	92.13	10	103.90	40	115.54	70	127.07
-15	94.10	15	105.85	45	117.47	75	128.98
-10	96.07	20	107.79	50	119.40	80	130.89
-5	98.04	25	109.73	55	121.32	85	132.80
0	100.00	30	111.67	60	123.24	90	134.70

Values in table are typical. Refer to specific manufactures data for exact values.

Figure 4.7-2 *100 Ohm Platinum RTD Response Table*

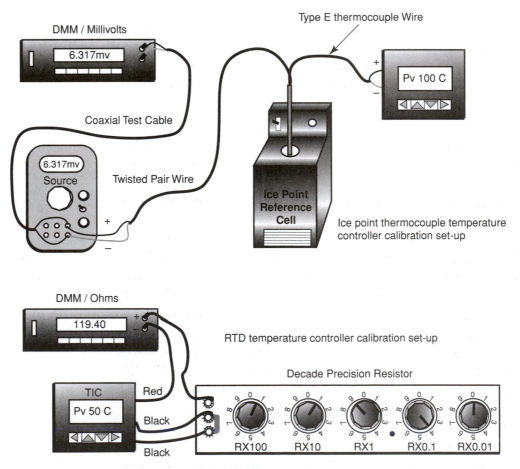

Figure 4.7-3 *RTD/Thermocouple Calibration Set-up*

4.8 Foreign Material Exclusion

Use care not to introduce foreign material or debris into the process during calibration activities. Clean permanent process fittings prior to opening, and thoroughly inspect all test fittings and tubing. If any uncertainty exists concerning the previous use of test tubing, clean and flush, or replace it. It is a good practice to use color-coded tubing and to paint all test fittings. Never introduce an unknown product into a process device. Always cover any openings into the process. During initial start-up, refuel cycles, or turn arounds, work often takes place inside vessels, tanks, condensers, and reactors. Every item entering these areas is inventoried at entrance and exit. Take only the necessary items into these areas. Attach lanyards to tools, and carefully control any debris generated at the location. Place exclusion covers or temporary caps on any opened piping.

■ Chapter Four Summary

- Electronic testing devices detect and measure basic characteristics of electrical signals such as frequency, period, pulse width, volts, amperes, resistance, rise time, decay time, and noise.
- Multimeters placed in the voltage function exhibit very high impedance and can usually measure voltage without disrupting the circuit. Multimeters in the ampere or ohm function act as a direct short and can cause serious damage if improperly installed.
- Precision resistors and decades are most commonly used to convert a current signal into a voltage and to simulate an RTD.
- Megohmmeters measure insulation breakdown voltage and are commonly used to detect faulty cabling. Do not attempt to megohm cables unless they are disconnected from all circuits. Ensure the megohmmeter is drained prior to disconnecting from a recently tested cable.
- Clamp-on ammeters are used for non-intrusive current readings and are useful to detect grounding faults.
- The oscilloscope monitors voltage in respect to time or compares two voltages. Oscilloscopes display a waveform of the monitored signal. Required value units are extracted from the waveform display and multiplied by the oscilloscope dial settings. The oscilloscope display reveals such signal characteristics as frequency, period, pulse width, voltage ACp-p, volts AC RMS, volts DC offset, rise time, decay time, and noise.
- All electrical and electronic testing and measurement activity requires particular awareness of isolations, grounds, and static discharge potentials.
- The electrical potential of a floating ground is substantially different from adjacent earth grounds. Completely isolate any test equipment used on such systems from plant common ground or earth ground.
- Common isolation transformers that use a three-prong outlet do not offer ground isolation.
- Electrostatic discharge can also cause irreparable damage to electronic circuits; always use anti-static mats and wrist straps when working with static sensitive components.

- Pressure testing equipment includes deadweight testers, test gauges, manometers, digital pressure detectors, hand pumps, and wet-leg calibrators.

- Since the fluid in the tester is applied to the device tested, ensure compatibility of the testing fluid to the process media, and/or flush the device prior to installation.

- Test gauges are similar to process gauges, the primary difference being a more precise manufacturing process and a mirrored anti-parallax dial face. Determine the correct resolution and accuracy prior to use, and exercise caution not to cross contaminate the process with residual fluids left in the gauge from previous use. Most test gauges are sensitive to orientation and require zero adjustment if repositioned.

- Manometers are used to measure low-pressure values. They are more commonly used as a bench-testing device.

- Digital pressure testers are commonly used to measure low-pressure values in the field. They are extremely accurate and more convenient to use than manometers. Allow sufficient warm-up prior to use, ensure the reference vent is open, do not use fluid as a calibration media, keep the internal transducer clean and dry, and do not over range the calibrator.

- Hand pumps are often used to develop the test signal input pressure. Use caution not to introduce contaminates into the system from the hand pump.

- Wet-leg calibrators are often used to calibrate boiler feed and level transmitters in the field. Wet-leg calibrators are a dual chamber fluid reservoir with separate high side and low side pressure inlet and outlet ports. The calibrator allows wet calibrations while keeping the hand pump and test gauge isolated from the process fluid. Since the high side and low side are both filled, the tubing length and elevation is self-canceling.

- It is important to continually agitate a fluidized temperature bath to obtain consistent values. Submerge the probe to an appropriate depth, but do not allow the probe tip to contact the wall or bottom of the bath. Allow sufficient temperature stabilization time prior to recording data.

- Ice point cells are often used to provide a reference junction when injecting voltage to simulate a thermocouple.

- It is a good practice to erect an exclusion barrier around a temperature bath to prevent inadvertent or unauthorized exposure.

- High temperature fluids can react violently to the introduction of other fluids.

- Never introduce an unknown product into a process device. Always cover any openings into the process. Attach lanyards to tools, and carefully control any debris generated at the location. Place exclusion covers or temporary caps on any opened piping.

Chapter Four Review Questions:

1. List the common testing devices used to detect and identify electrical and electronic signal characteristics.

2. List the common electrical signal characteristics identified and manipulated during troubleshooting or calibration activities.

3. State the precautions required when connecting a digital multimeter into a circuit to monitor amperes or ohms.

4. Explain the "live, dead, live" safety practice.

5. Determine the appropriate testing equipment required to measure 4 – 20 mADC as 1 – 5 VDC.

6. State the appropriate grounding precautions when installing a recorder.

7. Determine the appropriate test equipment to detect degraded instrument signal cabling.

8. State several uses and advantages of clamp-on ammeters.

9. Compare and contrast floating ground systems and conventional earth ground systems.

10. State several precautions required when working with or handling static sensitive components.

11. List several testing devices used to detect, simulate, or measure pressure.

12. State the adverse affects of over pressurizing test equipment and process devices.

13. State the precautions regarding contamination when using pressure testing devices.

14. Explain the proper selection, use, and operation of a test gauge.

15. Explain the benefits and limitations of using a wet-leg calibrator.

16. State the purpose of an ice-point reference cell.

17. Given access to a thermocouple table determine the millivolt equivalent for various temperature values.

18. Given access to an RTD table determine the ohmic equivalent for several temperature values.

19. State several safety considerations and precautions used when performing temperature calibrations.

20. State several precautions to prevent introduction of incompatible fluids into the process.

Chapter 5

Transducers and Transmitters

OBJECTIVES

Upon completion of Chapter Five the student will be able to:

- Describe the operation of a pneumatic transmitter, including the purpose of the input sensor, the force balance mechanism, and the pneumatic output relay.
- Describe several tests that determine transmitter operability.
- State the location of zero and span adjustments on a pneumatic transmitter.
- Describe the operation of an electronic transmitter, including the purpose of the input sensor, the high impedance inputs, the error detector, the amplifier/driver, and the isolated outputs.
- State the location of zero and span adjustments on an electronic transmitter.
- Describe transmitter calibration process when using a HART communicator.
- Describe the operation of a solenoid.
- Describe the operation of an electrical switching relay.
- Describe several tests that determine relay operability.
- State the hazards and precautions required when testing relays.
- State the operation and list the major functional components of an lvdt.
- Describe several tests that determine lvdt operability.
- State the operation of an eddy current proximeter (mil gap/VDC).
- Describe several tests that determine proximeter operability.
- State the gap voltage installation requirements of a proximeter.

- *State the operation of a seismic detector (G/volts).*
- *Describe several tests that determine seismic detector operability.*
- *State Ohm's law of equilibrium.*
- *State Kirchoff's law of conservation.*
- *Calculate the combined resistance of a series circuit.*
- *Calculate the combined resistance of a parallel circuit.*
- *Calculate the combined resistance of a series/parallel circuit.*
- *Develop a diagram of a simple bridge circuit.*

5.0 Introduction

A transducer is a device that converts one form of energy data into another proportionally equivalent form of signal data. A transmitter is a device that conditions and conveys signal data. These terms are practically synonymous as most modern transducers function as transmitters and most modern transmitters contain integral transducers. Transmitters and transducers are identified by the characteristics of the signal they transmit. The function of the transmitter is dependent on the installation and particular primary element monitored. A pneumatic transmitter with high and low vessel taps functions as a level transmitter. The same pneumatic transmitter upstream and downstream orifice taps functions as a flow transmitter. In either case the level or flow value detected is transmitted as a proportional and equivalent pneumatic signal. If an electronic transmitter is installed to detect level or flow, the same level or flow value is detected. The transmission data, however, is a proportional and equivalent electronic signal.

5.1 Transmitters

Pneumatic transmitters detect and convert a change in pressure or level into equivalent pneumatic signals and retransmit the pneumatic signal data. The most common pneumatic signal is 3 – 15 psig. Pneumatic transmitters usually incorporate a capsule assembly as the input sensor. The capsule converts any change in input pressure into physical displacement or motion. Capsule displacement is transferred to a force bar. The bar typically is attached to the capsule at one end and a balance beam at the other. The bar levers on a pressure-isolating fulcrum. A flapper nozzle assembly on the balance beam detects displacement of the bar as an error signal, see Figure 5.1-1.

The transmitter is supplied with 20 psig of continuous instrument air (supply). The measured variable value from the primary element is applied to the input sensor (input). The input sensor transfers an equivalent amount of energy to a force balance mechanism. A flapper nozzle assembly detects the amount of deviation at the force balance beam, see Figures 5.1-2 and 5.1-3.

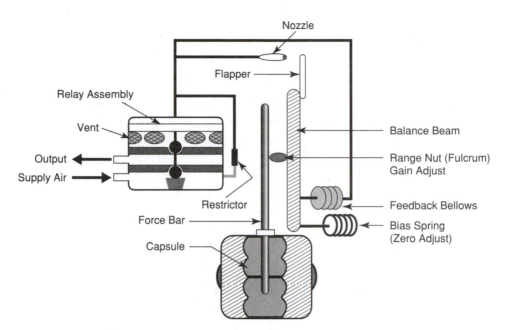

Figure 5.1-1 *Pneumatic D/P Transmitter*

The flapper nozzle assembly backpressure is amplified by a pneumatic relay and retransmitted as a pneumatic output signal. Figures 5.1-4, 5.1-5, and 5.1-6 depict several operational conditions. Supply air flows through a restriction screw and is supplied in parallel to the nozzle, feedback bellows, and upper chamber of the output relay. The flapper

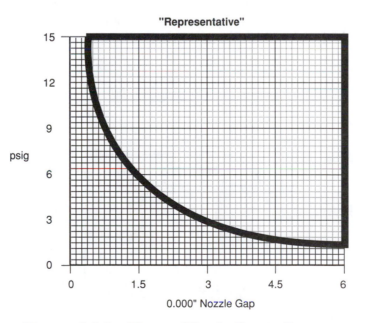

Figure 5.1-2 *Flapper/Nozzle Gap to Pressure*

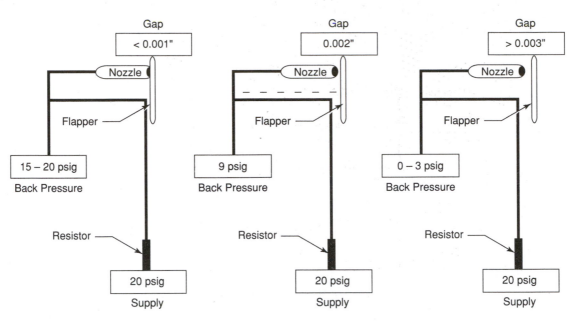

Figure 5.1-3 *Flapper/Nozzle Diagram*

acts to block the nozzle and build backpressure (error signal). The feedback bellows and output relay simultaneously sense the error signal. The feedback bellows expand to restore balance to the beam. The relay amplifies the error signal and outputs a representative value of supply air.

Flexible metallic strips (flexures) mounted parallel to the force bar suspend the balance beam. The flapper/nozzle is at one end of the balance beam and a feedback bellows and bias adjustment spring is at the other. Bias spring tension determines the initial balance

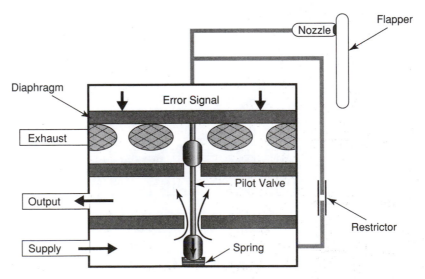

Figure 5.1-4 *Relay Output*

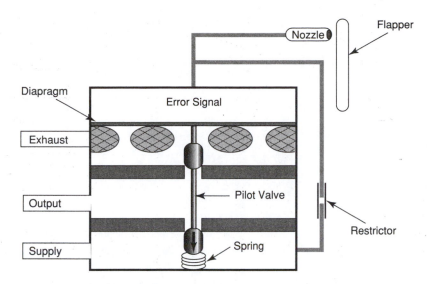

Figure 5.1-5 *Relay Output*

condition of the balance beam. The beam pivots on an adjustable range nut. Since range nut position determines the fulcrum point, shifting the axial position of the range nut relative to the balance beam determines the mechanical gain or multiplication of force bar motion to balance beam motion, Figure 5.1-7.

Operability Assessment: Pneumatic transmitters require clean, dry filtered instrument air. Most operability issues are related to air quality. Moisture at the nozzle or an erratic hissing sound often indicates poor air quality. Blow down and adjust the filter regulator. Transmitter relay assemblies have an integral clean out plunger or removable restrictor screw. Depress the plunger and run a fine wire through the restrictor assembly. Manually

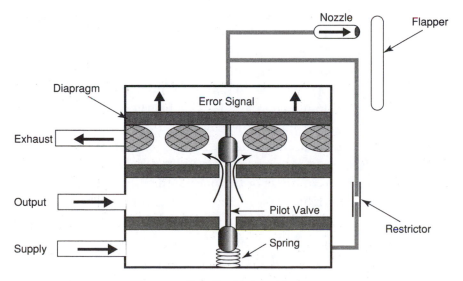

Figure 5.1-6 *Relay Output*

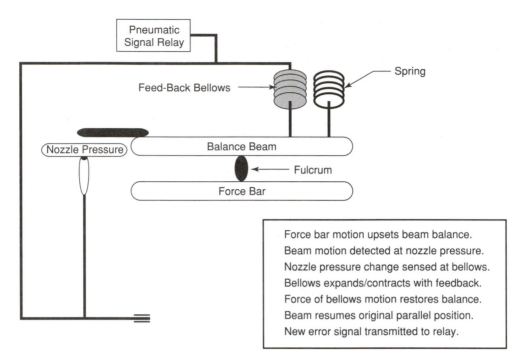

Figure 5.1-7 *Balance Beam Assembly*

depress the flapper against the nozzle and observe full supply pressure at the transmitter output. If the output does not increase to the supply value, the flapper/nozzle is worn or the relay is plugged. Gently pull the flapper away from the nozzle and observe the output drop to zero. If the output remains high, the nozzle is probably plugged and requires replacement. Check the feedback/output bellows for leakage.

Calibration Techniques: Unless specifically stated contrary, all differential device calibrations assume the measured input is applied to the high side and the low side is vented to atmosphere. Apply the lower range value and turn the zero spring screw to obtain zero output, typically 3 psig. Always ensure the final adjustment to a springed mechanism is in the direction that increases spring tension. Adjustments that reduce spring tension may relax over time and introduce error. This adjustment is usually accessible through a slot in the cover. Apply the upper range value and observe the correct output, typically 15 psig. If adjustment is required, remove the cover and adjust the range nut to obtain the desired value. Move the range nut closer to the flapper end of the range bar to decrease the output value; move the range nut away from the flapper end of the range bar to increase the output value. Always count the number of turns to determine sensitivity and track changes. Differential devices are susceptible to static zero shift. Introduce full static pressure value to both sides simultaneously and monitor for a change in output. Static alignment adjustments are provided by many pneumatic transmitters but are beyond the scope of this text; refer to the manufacturer's literature.

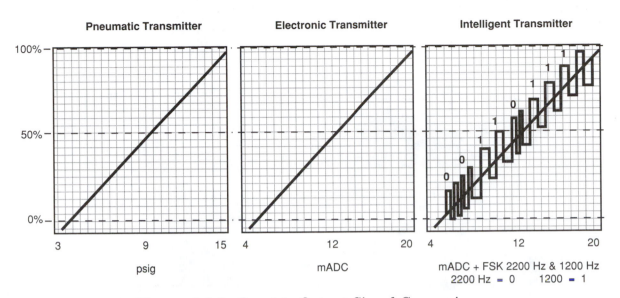

Figure 5.1-8 *Input to Output Signal Comparisons*

Electronic transmitters perform the same basic functions as pneumatic transmitters. The significant difference is the replacement of the force/balance mechanism with electronic components and the generation of an electrical signal output. Pneumatic transmitters output an air signal of 3 – 15 psig. Electronic transmitters produce an electrical output signal of 4 – 20 milliamps DC. HART compatible transmitters produce an electrical output of four to twenty milliamps DC with a frequency shift key data transmission signal superimposed. The frequency signal oscillates equally above and below the analog signal and has no effect on the total current value, as shown in figure 5.1-8. The most common electronic signals are 4 – 20 milliamps DC, 10-50 milliamps DC, and 1 – 5 volts DC. Zero and span screws accessed through the transmitter housing provide adjustment capability for standard electronic transmitters, Figure 5.1-9.

Electronic transmitters require an external power supply to provide 24 to 48 VDC. The transmitter is connected to the power supply with two wires. The two-wire connection provides power to operate the transmitter and also provides the propagation path for signal transmission. The transmitter varies the current flowing through the power/signal loop between 4 and 20 milliamps in relation to the process variable monitored. Transmitter circuitry consists of an input sensor (detector), connection board, calibration board, and an isolated output amplifier driver board. The detector converts physical displacement into a change of impedance or capacitance. The connection board accepts the ribbon cable from the detector and conditions the signal for the calibration board. The calibration board consists of high impedance operational amplifier error detector circuits with tunable resistors (potentiometers). High impedance detection circuitry prevents the error detection amplifier from placing a load (drawing current or dropping voltage) on the initial signal. The potentiometers are adjusted with the zero and span trim screws to correct calibration.

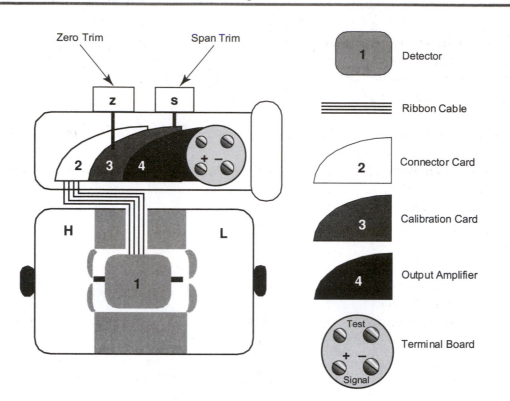

Figure 5.1-9 *Electronic Transmitter Diagram*

The isolating output driver board converts the internal signal into a modulated current signal on the power loop wiring, Figure 5.1-10. The isolated output limits current on the loop and prevents the amplifier from damage in the event of a signal loop grounding condition.

> **NOTE** *Depressing the zero and span buttons on a "Smart" transmitter will not result in a true and complete calibration. A communicator must be used to perform a sensor trim.*

Calibration Technique: To calibrate, apply the lower range value to the transmitter and observe zero indication, typically 4 mADC. Adjust zero screw as required to obtain correct

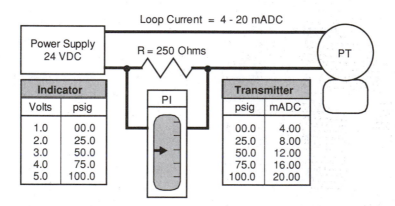

Figure 5.1-10 *Electronic Transmitter Signal Loop*

output value. Apply the upper range value and observe the upper output value, typically 20 mADC. Adjust span screw as required to obtain correct output value. Repeat as required to obtain satisfactory readings. Transmitter output is monitored as current at the transmitter test jacks, or as voltage across a series installed 250-ohm precision resistor.

 It is important to realize that even though the transmitter signal is 4 – 20 mADC, and is sensed as true milliamps when monitored, the transmitter is not actually providing current. The current is provided by (originates at) the instrument rack power supply; the transmitter merely modulates the amount of current allowed to flow through the signal path. When simulating a transmitter during loop calibrations, configure the signal source as "transmitter simulate," XMT SIM, not milliamps output.

Modern electronic transmitters are microprocessor based and have "intelligent" capabilities, Figure 5.1-12. They also require an external power source and modulate the loop current in relation to the process variable monitored. HART compatible transmitters monitor their own output signal loop for communication data. The binary communication data is superimposed on the current signal as an AC frequency. Two distinct frequencies are used to represent the binary states. Typically, 2200 hertz represents a logic state (0) and 1200 hertz represents a logic state (1). The AC frequency does not affect the current signal value, as it modulates equally above and below the current level and self-cancels. Each transmitter contains a packet of information called tag data. Tag data includes the polling address, manufacturer's specification, calibration data, and tuning algorithms unique to the particular device. Several transmitters often share a single two-wire power cable. This facilitates multidrop and bus-type control arrangements where data from many transmitters is continuously monitored, displayed, and recorded by a central control station. Handheld communicators and laptop computers access the intelligent transmitter software through the signal loop line to configure and calibrate the transmitter, Figure 5.1-13. Polarity is not critical with a handheld communicator. The loop must have at least 250 ohms resistance. Most smart or intelligent transmitters have security and fail mode jumpers. Place the security jumper in the "off" position to access the complete configuration menu of the

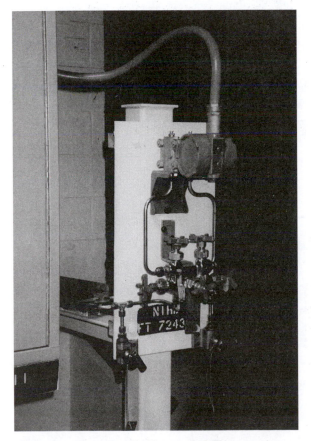

Figure 5.1-11 *Differential Transmitter*

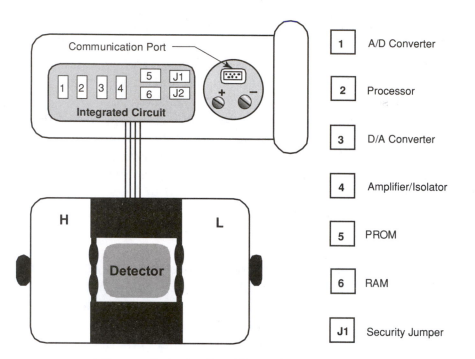

Figure 5.1-12 *Intelligent Transmitter*

transmitter. Some transmitters will only communicate with the security jumper in "off" position. The fail mode jumper determines whether the transmitter will fail with a high output or a low output. Traditional output range is 3.5 mADC to 20.5 mADC. Fail range high is typically >21 mADC; fail range low is typically < 3.5 mADC.

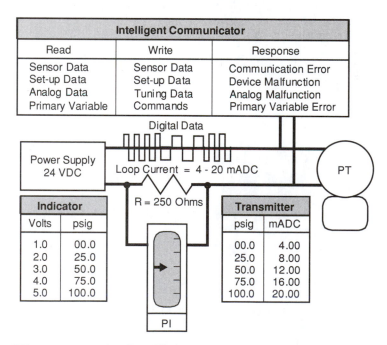

Figure 5.1-13 *Intelligent Transmitter Signal Loop*

Calibration Technique: HART transmitters require a communicator for complete calibration. The zero and span buttons effectively re-range the transmitter and only simulate a calibration. Connect the communicator to the two-wire loop. Connect a calibrated multimeter in ampere mode to the transmitter test jacks. Connect a variable calibrated pressure source to the input port of the transmitter. When energized, the communicator will search for the device. Navigate menu through set-up/HART output/ then select Write protect, and toggle off. Remove security jumpers if installed. Navigate menu through Set-up/Basic/ then select Damping, and remove. Navigate menu through Set-up/ Basic/ then select LRV, URV, apply pressure values and Send. Navigate menu through Service; then select Sensor trim, apply pressure values, and Send. Navigate menu through Service then select Output trim. Apply the desired lower output value and upper output value, 4 – 20 mADC and Send. Restore Write protect and security jumpers.

> **NOTE** *While polarity is not critical with hand-held calibrators and laptop computers, ground isolation is critical. Ensure proper ground isolation prior to connecting test equipment.*

When operating in the battery mode there is little if any grounding concern. It has been noticed at some facilities that isolation transformers often have a case ground. Laptop computers used with isolation transformers may actually still ground through the ground prong. Providing a ground path to either side of a current loop will take the loop value to zero and may cause unexpected perturbations. Even with the ground prong removed, the

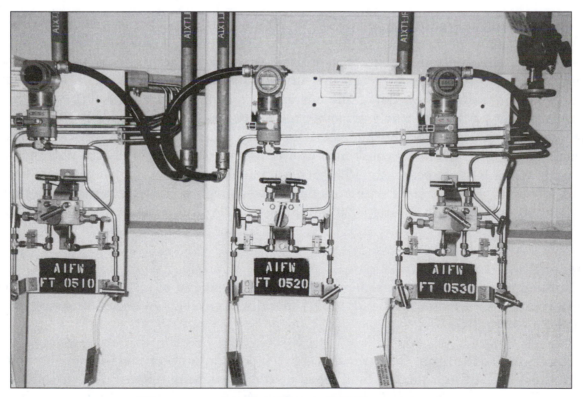

Figure 5.1-14 *Intelligent D/P Transmitters*

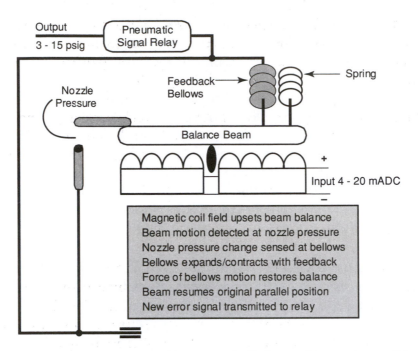

Figure 5.1-15 *Electro/Pneumatic Balance Beam Assembly*

isolation transformer may still case ground if placed on metal or concrete. It is not safe work practice to remove the ground plug. Recharge or replace the laptop batteries before field use.

Pneumatic transducers (I/Ps) and (E/Ps) convert electronic signals into pneumatic signals. They contain many of the same type components as pneumatic and electronic transmitters. The principal difference is, unlike transmitters, which detect a physical change in process energy, transducers merely convert one form of signal transmission data into another form of signal transmission data. The current to pneumatic transducer (I/P) contains a pneumatic relay, a flapper/nozzle assembly, a balance beam, a feedback bellows, and a coil. The operating principle of the pneumatic section is identical to that of the pneumatic transmitter. While the pneumatic transmitter relies on a force bar to initiate the unbalance condition to the balance beam, the I/P depends on an electrical coil to disrupt beam balance in proportion to the electrical current applied to it, see Figure 5.1-14. Typical current inputs are 10 – 50 milliamps DC and 4 – 20 milliamps DC. Typical outputs are 3 – 15 psig and 6 – 30 psig.

Operability Assessment: As with any flapper/nozzle assembly, careful manual manipulation can determine basic operability. Blow down and adjust the filter regulator to verify appropriate air supply. The I/P relay generally has a cleanout restrictor screw; remove and clean as required.

Calibration Technique: I/P transducers require calibration in their field-mounted orientation. Record and lift the field leads at the I/P terminal board. Inject a monitored 4-20 mADC current signal and observe the respective 3 – 15 psig output at a calibrated device. Allow sufficient warm-up time and exercise the coil for several cycles. It is recommended as with all regulated pneumatic outputs to induce a slight bleed during calibration and observe for droop. Adjust the zero screw and span screw as required to obtain correct output values.

Unexpected I/P failures can occur due to loose range configuration jumpers. Since most I/P installations control final device position, I/P failure typically results in control valve failure. Jumpers installed in particularly high heat and/or high vibration areas may require soldering to hold in place.

5.2 Physical Displacement Transducers

Physical displacement transducers convert physical displacement (motion) into an electronic signal or an electric signal into physical displacement. The fundamentals of solenoid, coil, and core behavior are observed as two magnets placed poles opposed create a repelling force between them sufficient to generate a physical displacement. Figure 5.2-1 illustrates a ferrous material placed in a magnetic field increasing the field strength, and a magnetic field applied to a ferrous material inducing a force. A complete explanation of electro/mechanical relationships controlling coil and core dynamics is beyond the scope of this text. Exact detail of these phenomena is found in the works of Michael Faraday and Emil Lenz.

Common physical displacement transducers include the solenoid, lvdt (linear variable displacement transformer), the eddy current detector, the seismic detector, and the velocity transducer. Solenoids convert electric current into physical displacement. They consist of a primary coil winding and a movable core (slug). The core is often attached to small three-way valves, but is also connected to strikers and plungers. Solenoids attached to small valves are often used to direct fluid for control purposes. Solenoid valves are also installed in the instrument air supply line to air operated valves to open or close the valve in response to process conditions. Solenoids attached to strikers and plungers lock and unlock doors,

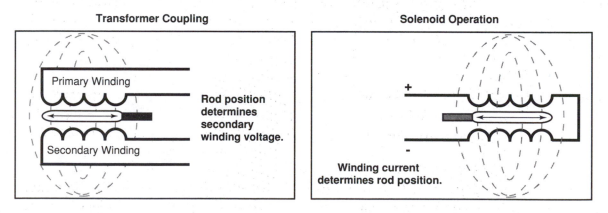

Figure 5.2-1 *Solenoid, Coil, Core Operation*

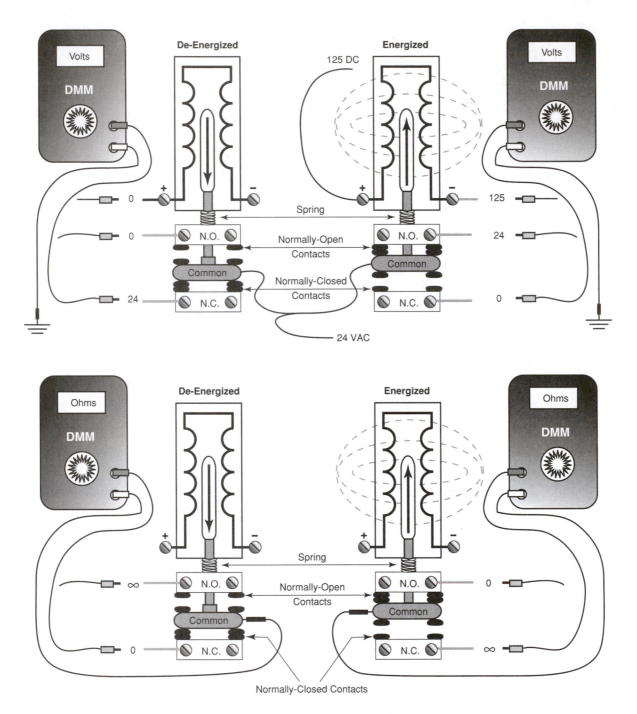

Figure 5.2-2 *Relay Solenoid Assembly*

push small items from conveyors, and perform a multitude of other functions where a small displacement device is electrically controlled. A relay is a set of switching contacts manufactured with an integral internal solenoid. The configuration depicted in Figure 5.2-2 utilizes the relay coil (solenoid) to pull in the switch contacts. Relays are used throughout industry for remote automatic switching of electrical devices. Solenoids and relays are the main control initiating devices of program logic functions. High voltage may be present at relay contacts. Use extreme caution when taking measurements. Relay cabinets often represent a reasonable "half-split" for troubleshooting. The control voltage leads to the coil generally come from the control system, such as a plc. The wetting voltage leads at the contacts generally operate field devices. It is good practice to first determine the signal present at the relay coil and contacts. Verify a multimeter by reading a known voltage source before attempting to take readings. Check the relay with one lead to ground and the other set to measure AC, measure all contacts and the coil. Repeat this process in DC mode. Then re-verify the meter to the known source. Since a multimeter in the ohms position induces a direct short, it is not advised to monitor contacts for ohms unless all disconnects and potential back feeds are determined and tagged out.

Relay position is sometimes determined through visual inspection. An energized relay is "pulled in." When de-energized a small plunger will extend through the center of the relay cover. "Force" the relay by depressing the plunger manually, and test the contacts for good open and closed indications. Energized contacts that are closed will both read full voltage to ground but zero voltage one to the other. Energized contacts that are open will read full voltage to ground from one contact and no voltage to ground from the other. Open contacts read full voltage one to the other.

The linear variable displacement transformer (lvdt) converts linear displacement into an equivalent change in voltage. Their most common application is as position feedback transmitters. The lvdt is installed on a valve stem or mechanical lever. Any change in the position of the valve stem or lever is transmitted back to a control system such as a servo control assembly. The servo control assembly compares desired position to actual position and develops a correction signal to correct any deviation. The lvdt assembly consists of an excitation oscillator, a primary transformer, a secondary transformer, and a movable core. The excitation oscillator generates an alternating voltage to the primary transformer. The primary transformer generates a flux field to the secondary transformer. The secondary transformer consists of two separate coils wired series opposing. The position of the movable core determines the amount of coupling strength to the secondary coils. As the secondary transformer is series opposed, and increasing coupling to one coil decreases coupling to the other, voltages range from positive polarity saturation to negative polarity saturation with the exact center self-canceling. Figures 5.2-3, 5.2-4, and 5.2-5 show each position. The center travel position where the voltages sum to zero is the null position. Lvdts are produced to detect a wide range of values. The outputs are always bipolar (going positive and negative) with zero volts at the null or center position.

Operability Assessment: Lvdts are permanently damaged if wired incorrectly. The preferred method to test an lvdt is with it connected to the driver device. Determine the output leads and measure for volts DC. Manually extend the lvdt and observe the voltage output change. Position the lvdt in the opposite direction and observe voltage of equal magnitude but reversed polarity. The preferred installation alignment mechanically centers the lvdt at the electrical center, null, or zero volts.

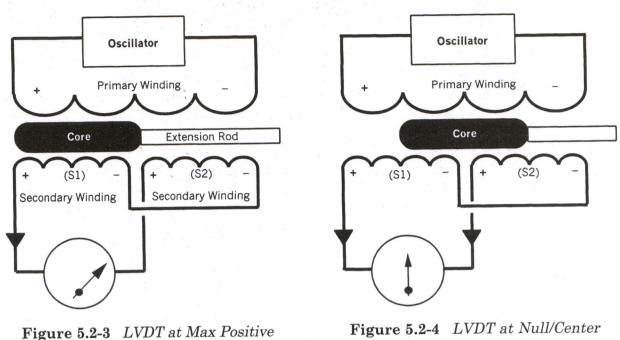

Figure 5.2-3 *LVDT at Max Positive Position*

Figure 5.2-4 *LVDT at Null/Center Position*

Eddy current detectors convert mils gap into voltage. Mils gap is the distance between the probe tip and the metallic object monitored measured in one-one thousandths of an inch, Figures 5.2-6 and 5.2-7. An excitation oscillator provides a high frequency voltage to the detector coil. The energized coil emits a magnetic field. The strength of the magnetic field changes as metallic mass passes through close proximity. The change in field strength is detected and converted into an equivalent voltage value (-0.200 VDC per mil). Eddy detectors are most often used as non-contacting vibration detectors on rotating equipment. Typical input range values are 0-100 mils producing outputs of 0 to (-) 20 VDC. Proximity probes, proximeters, and the connecting cable are impedance tuned as a set. Any single component replaced must match the tuned electrical length of the original component. Probe installation gap voltage is determined by the clearances required at the probe tip. A gap voltage of (-) 10 volts indicates 0.050" clearance between the probe tip and the object monitored. Non-conducting gap feeler gauge sets are available from the manufacturer.

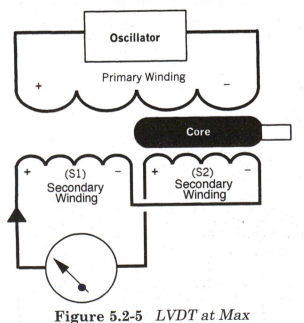

Figure 5.2-5 *LVDT at Max NegativePosition*

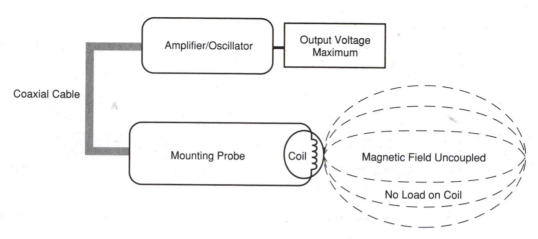

Figure 5.2-6 *Proximity Detector*

Calibration Technique: Dynamic calibration of proximity probes requires a wobulator table. Proximity probes however, are often static tested in a head micrometer with good result. Ensure the proximeter is energized and connected to the probe. Mount the probe perpendicular to the micrometer head with the probe tip just "not" contacting the metal surface of the micrometer. Output voltage at the proximeter's signal and common terminals should read less than two volts. As the micrometer head is dialed away from the probe tip, output voltage should increase approximately (-) 0.200 volts per mil (1/1000") until the proximeter reaches full output value, (-) 18 to (-) 24 volts.

The threaded cable connectors are part of the shield and require electrical isolation from ground. Spurious transients and spikes on a vibration channel can plague a facility. Cable connectors intermittently contacting an earth grounded metallic flexible conduit result in ground loop induced spikes on the signal cable. Most manufacturers provide a boot kit to cover the connectors.

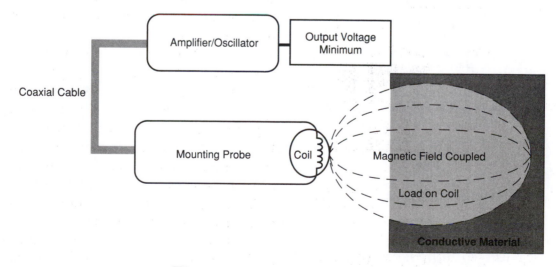

Figure 5.2-7 *Proximity Detector*

Seismic detectors convert gravity into volts. The components of seismic detectors include a strain gauge or piezoelectric crystal and an amplifier. Seismic detectors are used to detect ultra high frequency vibration. Velocity detectors convert acceleration into volts. A movable suspended core inside an excited coil detects velocity. An amplifier provides the excitation current to the coil. The suspended core moves in response to accelerating forces. The output voltage is generated as the movable core cuts lines of coil flux. The rate at which the lines are cut determines the amplitude of the voltage generated.

Operability Assessment: Complete calibration requires a shaker table, however operability tests are often performed in the field. Seismic detectors are rotated through each plane to indicate (+/-) one G. Velocity detectors produce a spike output in response to mechanical agitation.

5.3 Electronic Transducers

Electronic transducers are generally not used as stand-alone instruments. They are usually integrated with other components to detect outputs of thermocouples, RTD, thermistors, mass flow detectors, and analytical cells. Investigation of electronic transducers requires the application of several fundamental laws of electricity. The common units are volts, amperes, and ohms. Voltage (E) refers to the potential energy available and is sometimes considered analogous to the pressure of a fluid. Amperes (current or I) refers to the actual quantity of electrons per second flowing past a fixed point. Resistance (ohms or R) is the amount of opposition to current flow exhibited by a circuit or device. Resistors in series circuits create an opposition force equivalent to the sum of the individual resistances. Resistors in parallel circuits exhibit an opposition force equivalent to the reciprocal of, the sum of the reciprocals of, the individual resistors. Examples are given in Figures 5.3-1 and 5.3-2.

Series Resistors

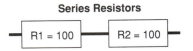

R total = R1 + R2 = 100 + 100 = 200 Ohms

Parallel Resistors

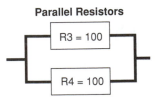

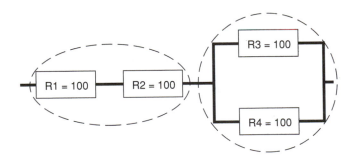

R total = 1/(1/R3 + 1/R4) = 1/(1/100 + 1/100)
R total = 1/(0.01 + 0.01) = 1/0.02 = 50

Figure 5.3-1 *Simple Series and Parallel Resistance*

R total = (R1 + R2) + [1/(1/R3 + 1/R4)]
R total = (200) + (50) = 250 Ohms

Figure 5.3-2 *Combination Series Parallel*

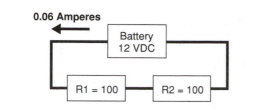

0.06 Amperes

Ohms equilibrium statement E/I × R = 1
Substitute knowns 12/I × 200 = 1
Solve for unknown 12/200 = I = 0.06 Amperes

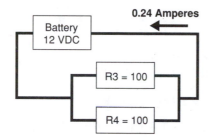

0.24 Amperes

Ohms equilibrium statement E/I × R = 1
Substitute knowns 12/I × 50 = 1
Solve for unknown 12/50 = I = 0.24 Amperes

Figure 5.3-3 *Simple Series/Parallel Circuit*

The first basic (Ohm's law) is actually a statement of equilibrium. The fundamental equilibrium of an electric circuit is stated as E/I × R = 1, or E = I × R (voltage is equal to the product of current and resistance). To determine the unknown values substitute the known values into the statement and reduce. The second principle is Kirchoff's statement of conservation. It states that the sum of the voltage drops across a circuit are equal to the total voltage applied to the circuit, and that the amount of current entering a circuit must equal the amount of current leaving the circuit. The third principle is Thevenin's theorem, which states that any combination of series and parallel components in a circuit can be represented by a single equivalent voltage and resistance, as in Figures 5.3-3 and 5.3-4.

There are many different applications of electronic detection transducers. Common elements include a bridge circuit and an amplifier section. The most basic designs also include a De'arsonval meter or galvanometer. The bridge circuit is the building block of micro-detection circuitry. The bridge is a configuration of parallel precision resistors arranged in such a manner as to detect the minutest changes in current between them. Once the bridge is balanced (nulled), the monitored component is placed into the circuit. Any deviations induced are attributed directly to changes in the monitored component. Figures 5.3-5, 5.3-6, and 5.3-7 depict three different conditions of bridge bias. Bridge circuits detect deviations in temperature, flow, and analytical cells.

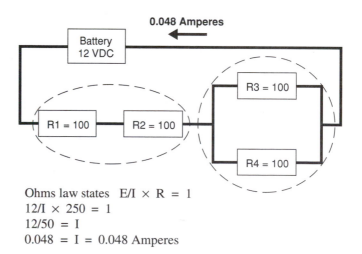

0.048 Amperes

Ohms law states E/I × R = 1
12/I × 250 = 1
12/50 = I
0.048 = I = 0.048 Amperes

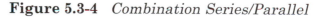

Figure 5.3-4 *Combination Series/Parallel*

119

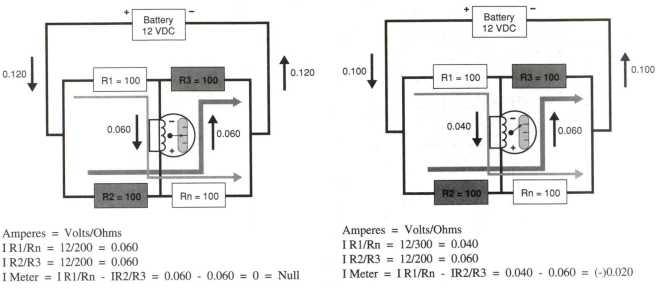

Amperes = Volts/Ohms
I R1/Rn = 12/200 = 0.060
I R2/R3 = 12/200 = 0.060
I Meter = I R1/Rn - IR2/R3 = 0.060 - 0.060 = 0 = Null

Figure 5.3-5 *Bridge Nulled for 100 Ohms*

Amperes = Volts/Ohms
I R1/Rn = 12/300 = 0.040
I R2/R3 = 12/200 = 0.060
I Meter = I R1/Rn - IR2/R3 = 0.040 - 0.060 = (-)0.020

Figure 5.3-6 *Bridge Detecting 200 Ohms*

5.4 Heat Trace

Heat tracing for transmitters and impulse lines prevents fluids in the lines from freezing and in some instances prevents particular process medias from cooling to a solid state. Steam and electric heat cable are used as heat tracing. Steam systems transfer heat to the sensing lines by providing steam flow through adjacent tubing. As steam relinquishes heat to the sense lines it cools and condenses. Condensation traps are installed at various low points in the steam tubing runs to collect moisture, Figure 5.4-1. Steam traps are designed to vent accumulated condensate when it reaches a predetermined level in the trap. Most steam trace performance problems are related to steam trap malfunctions, Figure 5.4-2. A drip pot that fails to blow down forces condensate back into the steam heat tubing and stops steam flow.

Electric heat trace systems use heat strip tape to transfer heat to sense lines. Heat tape is a semiconductive wire pair. It looks like a rigid extension cord and is produced in various lengths and power ratings. Various methods are used to energize the heat strip. More

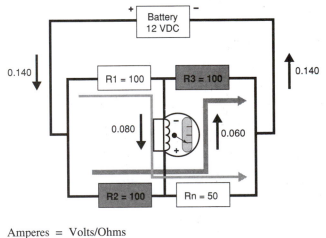

Amperes = Volts/Ohms
I R1/Rn = 12/150 = 0.080
I R2/R3 = 12/200 = 0.060
I Meter = I R1/Rn - I R2/R3 = 0.080 - 0.060 = 0.020

Figure 5.3-7 *Bridge Detecting 50 Ohms*

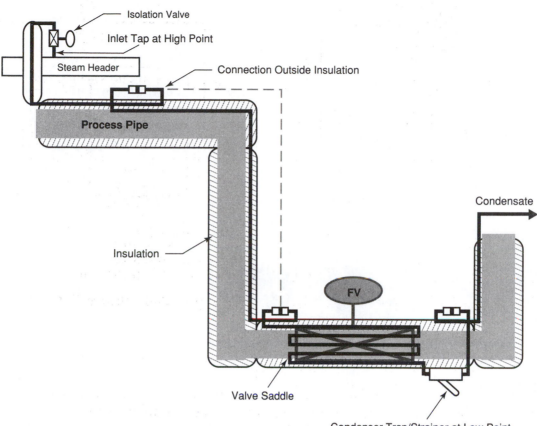

Figure 5.4-1 *Steam Trace*

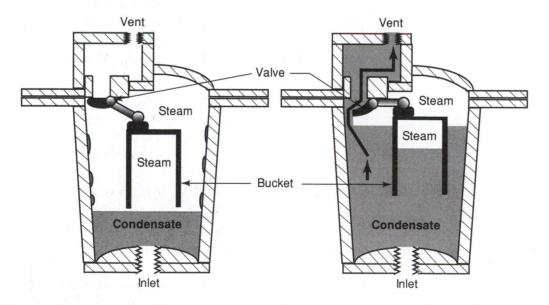

Figure 5.4-2 *Bucket Steam Trap*

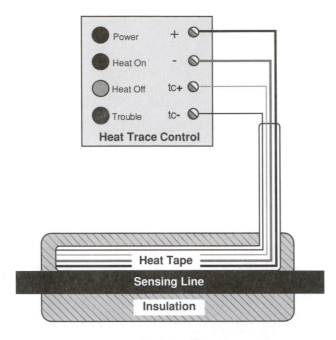

Figure 5.4-3 *Electric Heat Trace*

complex systems feature a temperature control unit that monitors the temperature of each individual strip and adjusts heating current as required to maintain a preset temperature, Figure 5.4-3. A separate module with status and diagnostic LEDs (light-emitting diodes) provides power and indication for each section of heat strip. Heat tape degrades with time and, as with any power supply, modules occasionally fail. Continuity testing generally determines the condition of the heat strip. Module failure is readily tested by component substitution.

Figure 5.4-4 *Electric Heat Trace Control Panel*

■ Chapter Five Summary

- A transducer is a device that converts one form of energy data into another proportionally equivalent form of signal data. A transmitter is a device that conditions and conveys signal data.
- Transmitters detect and convert changes in pressure, flow, or temperature into equivalent pneumatic or electronic signals and transmit the signal data from the field to a control center instrument (controller, recorder, or indicator).
- A flapper nozzle assembly detects the amount of deviation at the force balance beam. The flapper nozzle assembly backpressure is amplified by a pneumatic relay and retransmitted as a pneumatic output signal.
- Handheld communicators and laptop computers access the intelligent transmitter software through the signal loop line to configure and calibrate the transmitter.
- I/P and E/P transducers convert electrical signals into equivalent pneumatic signals. They are commonly used in the field to supply instrument air to final control elements.
- Physical displacement transducers convert physical displacement (motion) into an electronic signal or an electric signal into physical displacement.
- Common physical displacement transducers include the solenoid, lvdt (linear variable displacement transformer), eddy current detector, seismic detector, and velocity transducer.
- Solenoids convert electric current into physical displacement.
- A relay is a set of switching contacts manufactured with an integral internal solenoid. Relays are used throughout industry for remote automatic switching of electrical devices.
- Solenoids and relays are the main control initiating devices of program logic functions.
- The lvdt converts linear displacement into an equivalent change in voltage.
- Eddy current detectors convert mils gap into voltage.
- Electronic transducers are generally integrated with other components to detect outputs of thermocouples, RTDs, thermistors, mass flow detectors, and analytical cells.
- Resistors in series circuits create an opposition force equivalent to the sum of the individual resistances.
- Resistors in parallel circuits exhibit an opposition force equivalent to the reciprocal of, the sum of the reciprocals of, the individual resistors.
- Ohm's law states the fundamental balance in any electric circuit is stated as $E/I \times R = 1$.
- Kirchoff's statement of conservation is that the sum of the voltage drops across a circuit are equal to the total voltage applied to the circuit, and that the amount of current entering a circuit must equal the amount of current leaving the circuit.
- A Wheatstone bridge is a configuration of parallel precision resistors arranged in such a manner as to detect the minutest changes in current between them.
- A balanced bridge is nulled. Bridge circuits detect deviations in temperature, flow, and analytical cells.

Chapter Five Review Questions:

1. Define transducers and transmitters.

2. Describe the function of a pneumatic transmitter.

3. State the purpose of the input sensor.

4. State the purpose of the force balance mechanism.

5. State the purpose of the output relay.

6. Describe the function of an electronic transmitter.

7. State the purpose of the input sensor.

8. State the purpose of high impedance and isolation.

9. State the purpose of the error detector.

10. State the purpose of the amplifier/driver.

11. Contrast isolated and un-isolated outputs.

12. Describe the operation of a solenoid.

13. Describe the operation of an electrical-switching relay.

14. State the function of an lvdt (inches/VDC).

15. State the function of an eddy current proximeter (mil gap/VDC).

16. State the function of a seismic detector (G/volts).

17. Calculate the combined resistance of a series circuit.

18. Calculate the combined resistance of a parallel circuit.

19. Calculate the combined resistance of a series/parallel circuit.

20. Develop a diagram of a simple bridge circuit.

Chapter 6

Analyzers

- *State the adjustment used to correlate an analyzer reading to a known sample.*
- *Describe the troubleshooting technique used to isolate a reference temperature malfunction.*
- *State the purpose of chromatography as used in industrial applications.*
- *List three steps of the chromatographic process.*
- *State the purpose of a sampler valve and the purpose of the stationary packing.*
- *State the purpose of the spectrometer as it applies to chromatography.*
- *List and describe the four functional sections of a spectrometer.*
- *Explain the relationship between adsorption, travel time, and concentration.*
- *List several potential chromatograph malfunctions.*
- *Explain trace detection as used by chromatographs in control applications.*

6.0 Introduction

In most process plants analytical test are continuously performed to determine the quality of raw materials, intermediate, and finished products of the unit. Most of these tests are in the form of chemical analysis. Chemical analysis is the science and technology used to identify and measure the chemical composition of a product. Qualitative analysis determines the presence of a particular substance in a sample. A test to determine whether a sample contains CO_2 or that the sample has a certain pH is an example of qualitative analysis. Quantitative analysis determines the amount of a particular substance in a sample. Such determinations as the percent of oxygen or the "parts per million" of a contaminate are examples of quantitative analysis. Some analyzers, such as a gas chromatograph, perform both qualitative and quantitative analysis. Measured values are expressed as percent of concentration, parts per million (PPM), parts per billion (PPB), moles, seimens, micro mhos, and pH. Common industrial applications of chemical analysis include monitoring for product purity, verifying concentration of solutions, and detecting unwanted or undesirable contaminates. Industries maintain chemical laboratories to support skilled technicians and chemists that routinely pull and perform detailed laboratory analysis of product samples. On-stream analytical instruments are installed to perform continuous on-line monitoring, recording, control, and annunciation of critical phases of an operation. This helps to assure in-process product quality and reduces chances for environmental pollution. Most on-stream analyzers work on the same principles as their laboratory counterparts, except for the addition of mechanisms and circuitry required to support unattended automatic sampling. Process analyzers are classified in various ways depending on the purpose of the classification. Classifications are (a) by operating principle (infrared, ultraviolet, chromatographic, etc.), (b) by type of analysis (pH, oxygen, carbon dioxide, etc.), and (c) by selection (selective or non-selective). Infrared detectors are sensitized to monitor a single component while a chromatograph may monitor several components.

The number of different analyzers employed in the process industries makes up a significant list. They include pH analyzers, gas chromatographs, mass spectrometers, colorimeters, viscometers, density/specific gravity, sulfur analyzers, oxygen analyzers, opacity, conductivity, combustible gas, and turbidity. There are many others developed specifically for refinery streams such as ASTM distillation, pour point, flash point, and octane.

In some applications, analytical instruments monitor the chemical component of the process variable and provide control action. Analyzer control loops are common in chemical feed systems. Percent concentration of a particular chemical is monitored as the process variable. Sample concentration is compared to set-point, and the analytical controller outputs a demand signal to the chemical feed pumps that increases or decreases the amount of chemical injected with each stroke of the pump. Many processes are extremely sensitive to oxygen. Unstable compounds can exert tremendous energy if allowed to oxidize, and combustible processes can spontaneously ignite if exposed to oxygen. Oxygen analyzers are often installed to reduce corrosion and combustibility in these processes. The five analyzers reviewed in this chapter are pH, conductivity, oxygen, spectrometers, and gas chromatographs.

6.1 pH Analyzers

pH-monitoring analyzers measure the alkalinity or acidity of an aqueous solution. pH is actually a measurement of hydrogen ion concentration. Acidic solutions contain more hydrogen ions than base solutions. pH is the negative logarithm of hydrogen ion concentration. Therefore, a change of one unit pH is equivalent to a tenfold change in the effective strength of an acid or base. Highly acidic concentrations are referenced at (0) on the pH scale. Pure water is neutral (7), and highly alkaline (base) concentrations are referenced at (14). Figure 6.1-1 demonstrates the pH of several common products. Since pH measurement is temperature sensitive, pH analyzers contain a temperature-sensing element. pH probes are a combination of three detectors: a measurement cell, a reference cell, and a temperature compensation detector. The most common type of pH sensing element consists of a glass measurement electrode, a calomel reference electrode, and temperature compensation RTD, as shown in Figure 6.1-2. Measurement cells detect the

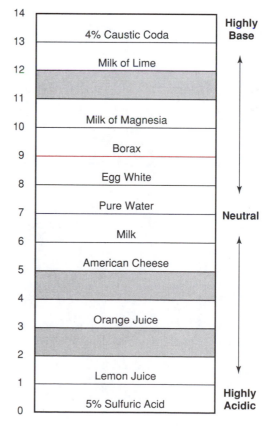

Figure 6.1-1 *The pH Scale*

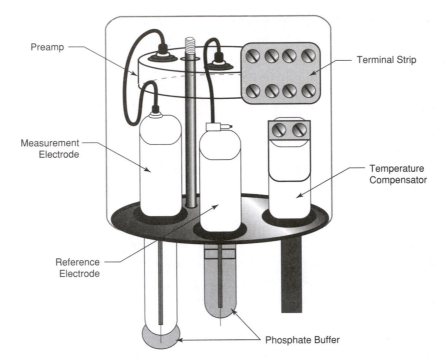

Preamp

Terminal Strip

Measurement
Electrode

Temperature
Compensator

Reference
Electrode

Phosphate Buffer

Figure 6.1-2 *Analytical Probe Assembly*

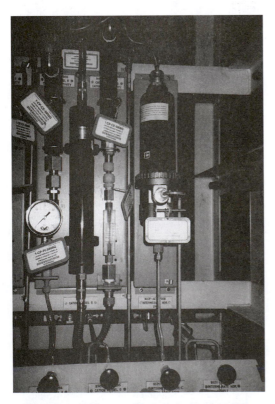

Figure 6.1-3 *pH Probe Assembly*

concentration or activity of the sample: reference cells generate a constant fixed reference voltage; and temperature compensators provide a temperature feedback signal. Measurement and reference cells function on principles similar to a battery. Conductive elements are selected to react with a specific electrolytic solution. As the conductor is placed in solution, a chemical reaction occurs. This reaction results in an ionic transfer between the solution and the conductor. Anode elements develop a positive charge and cathode elements develop a negative charge. The potential difference between charges is proportional to the logarithm of the analytic solution concentration. pH reactions are temperature sensitive and require accurate temperature sensing feedback. An RTD or thermistor circuit placed in the sample solution at the same location as the detection cells develops a temperature-related resistance value. Electronic circuits in the analyzer use the signal to develop a temperature compensation logarithm. This logarithm is applied to the cell signal to develop a temperature-corrected value. Probe-generated voltage values are inherently diminutive in magnitude. Bridge circuits located in the analyzer detect the minute changes in cell signal values. The signals are then amplified and processed as normal electronic data. Most pH analyzers provide local and remote indication, local and remote annunciation, and controlled output capability. Output indication is scaled as 0 – 14 pH. See Figure 6.1-3.

6.2 Oxygen Analyzers

Two common oxygen analyzers are the paramagnetic phototube detector and the galvanic probe detector. Paramagnetic analyzers detect the relative magnetic susceptibility of gasses. Oxygen has a strong attraction to magnetic fields while most other gasses have either a weak attraction or repulsion to magnetism. Figure 6.2-1 shows the magnetic susceptibility of several gases. Phototube detectors consist of a light source, a test body/mirror assembly, a beam splitting mirror, two phototube/optical amplifiers, a detecting bridge, and the supporting electronic amplifiers and signal conditioning circuits, see Figures 6.2-2 and 6.2-3. The test body/mirror assembly is suspended in the test chamber and subjected to a magnetic field. A light source is directed at the test mirror. The test mirror directs the light

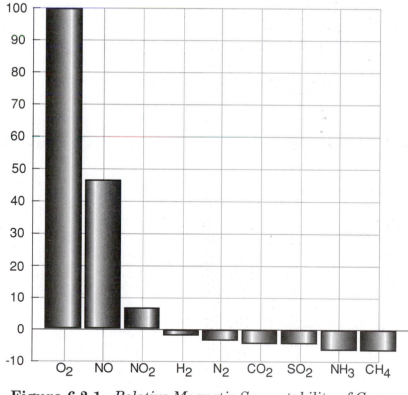

Figure 6.2-1 *Relative Magnetic Susceptability of Gases*

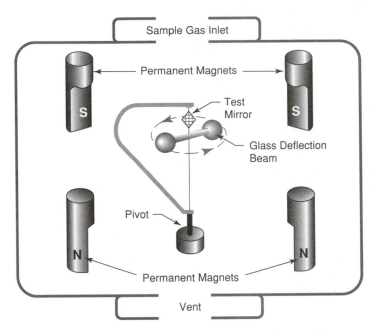

Figure 6.2-2 *Test Body/Mirror Assembly*

beam to a beam splitting mirror. Under null conditions, the beam splitter divides the light equally between the phototubes so their resultant voltage output is equal. As the sample gas is introduced into the test chamber, the test body deflects proportional to the magnetic susceptibility of the sample gas. The attached mirror rotates with the test body and redirects the light distribution to the beam splitter. Because phototubes generate a voltage output equivalent to the amount of light directed to them, the difference in output between the phototubes is proportional to the magnetically induced deflection of the detector mirror.

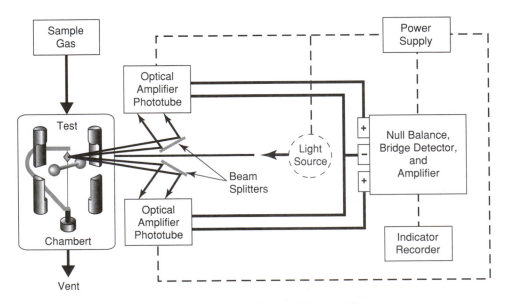

Figure 6.2-3 *Functional Block Diagram*

Figure 6.2-4 *Oxygen Probe with Auto Clean Cycle*

This difference is detected and amplified by the bridge and amplifier section and conditioned into a suitable readout. Operability checks are performed by admitting normal atmosphere into the test body and observing an indication of 20.7%. Figure 6.2-4 shows an oxygen probe with an automatic cleaning cycle.

Electrolytic probes often detect dissolved oxygen in fluids. Electrolytic oxygen analyzers use a platinum micro-cathode measuring electrode and a silver-silver chloride macro-anode reference electrode placed in a phosphate buffer solution. An oxygen permeable membrane isolates the probe from sample. Oxygen molecules cross the membrane and generate a small current. Electronic circuits in the analyzer bridge detect and amplify this current. Oxygen analyzer output indications are scaled as percent oxygen. Oxygen analyzer malfunctions include contamination and leaks into the sample line.

6.3 Conductivity Analyzers

Conductivity analyzers detect the ability of a sample to support electrical current. Conductivity analyzers are commonly used to detect ionized particles in pure water. A pair of gold or platinum elements is precisely positioned a fixed distance apart. Sample solution is routed between the elements. Oscillating voltage is applied to the elements. Current flow between the elements is representative of the number of ions available to carry current (conductance). Conductance (S) is the reciprocal of resistance. Therefore S = 1/ohm. The international unit of conductivity is siemens, although it is often referred to as a mho.

Example: Absolute pure water has a specific conductance of 0.055 micro mho at 25 degrees C.

To convert conductance into resistance divide the conductance into one. 1/0.055 micro mho = 18.18 Kohms. To obtain the actual expected value multiply the result (18.18 Kohm) by the cell constant of the probe.

If one cubic centimeter of pure water is measured for resistance with a volt/ohm meter and the meter probes are held exactly one centimeter apart and in perfect alignment the resistance will read 18.18 Kohms. If the water is exactly 25 degrees C, a platinum RTD temperature compensator will read 109.73 ohms. Extremely pure samples react more dramatically to ionic impurity than marginally contaminated samples react to additional impurity. It is also beneficial to determine the specific conductance of a sample. Specific conductance (K) is the conductance detected by one square centimeter probes through one cubic centimeter of sample. Conductivity cells are available with different degrees of sensitivity. Cell sensitivity is expressed as the cell constant. The cell constant is referenced to a standard solution of potassium chloride, and constant values are typically 1.0, 0.1, and 0.01. Analyzers are capable of using probes of any cell constant factor as long as the cell constant value is identified. Software programming or dipswitches inside the analyzer facilitate cell constant identification. Output indication values are expressed as micro mho or micro siemens. Figure 6.3-1 shows a conductivity sampling system. Four heat exchangers at the bottom of the panel stabilize the sample temperature. Eight valves in the center of the figure regulate sample flow rate. Four resin packed columns filter and stabilize the sample. The four analyzers monitor conductivity.

Figure 6.3-1 *Conductivity Sample System*

Operability Assessment: Modern probe analyzers are extremely accurate, but they do require proper maintenance. Erratic readings occur when sample flow rates or sample temperatures fluctuate excessively. Precise (repeatable) values are best obtained with stable sample flow rates and sample temperatures. Representative values are obtained when the probe mounting location and sample line installation provide a thoroughly mixed and contaminant free sample solution. Ensure the probes are clean and free of foreign debris. Manufacturers specification literature provides recommended sample flow rates and sample temperatures, and specifies proper probe cleaning techniques. The most reliable test for a probe analyzer is to compare the readings to a known good sample (standard). Many analyzers have a face-mounted adjustment to correct the reading to match the standard solution value; this adjustment is called standardization and is sufficient for most processes. A more accurate method is to obtain two different standard solutions, preferably one of a low value and the other of a high value. Rinse the probe with a neutral solution and place it in the low value standard. When the indication stabilizes, adjust the reference circuit to match the standard. Then, rinse the probe with a neutral solution and place it in the high value standard. Once the reading stabilizes, adjust the slope circuit to match the high value standard. If the sample delivery is within specification, the probes are clean, and the standardization procedure does not yield acceptable results, the probes or temperature compensator are most likely degraded. Prior to taking measurements, disconnect the probe leads from the analyzer. Temperature compensation resistance is measured with a digital multimeter in the ohms position. The ohm value obtained is referenced to an RTD response chart and compared to the known sample temperature. Cell resistance is measured and numerically inverted into conductance. Compare the measured conductance to the known sample conductance. If the values agree, the probe is functional. To test the analyzer circuits connect a precision resistance device adjusted for the correct ohm value to the analyzer and observe the analyzer reading. If the readings agree the analyzer is functional and the probes are suspect. If the analyzer and probe test good, the problem is sample delivery, probe placement, or electrical noise. Keep probes moist at all times. Do not use a replacement probe that has dried out.

6.4 Gas Chromatography

Mikhail Tswett first developed column chromatography as early as 1906. He was able to separate plant pigments by pouring an ether solution containing leaf material into a vertical glass column packed with calcium carbonate. As the solution migrated down through the column, different pigments stratified at different heights in the column. This resulted in a chromatogram, a series of horizontal bands of color along the height of the column with each band composed of a different pigment.

One of the most valuable analyzers in the process industry is the gas chromatograph. While most analyzers are instruments that directly monitor the product and provide instantaneous analysis, chromatography is actually an analytical process that samples, separates, detects, qualifies, and quantitates a product. The chromatograph consists of a programming controller section, a sample collection section, a separation section, and a detection section. Refer to block diagram Figure 6.4-1.

The program control section provides the logic sequence to automatically manipulate the sample and column valves. It contains the timer, the power supply for the bridge circuits, the automatic zero mechanism, adjustment pots, and signal conditioning circuits. Several automatic functions are switch selectable. Typical programs include chromatogram mode, calibration mode, or automatic analysis mode. Programmable temperature control and automatic stream switching is also supported. The output display device is usually a strip chart recorder.

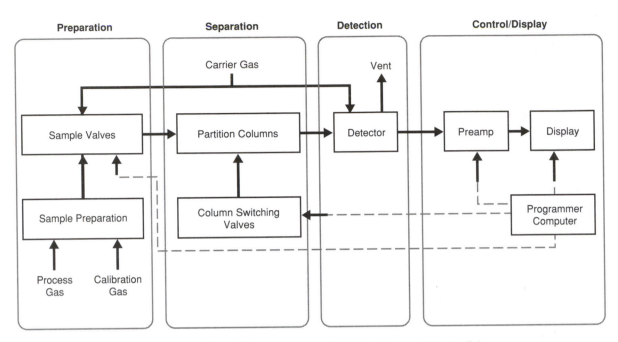

Figure 6.4-1 *Gas Chromatographic Analyzer Block Diagram*

The sample collection section consists of the piping, tubing, valves, filters, dryers, and heaters needed to collect, prepare, and inject the sample and other gases. Product samples are syringe injected directly into a septum on laboratory analyzers. On-line analyzers use sample valves to introduce a fixed volume sample into a carrier gas stream that transports the sample through the column. Multiposition valves alternately route purge gas, calibration gas, and sample gas through the column. The samples are filtered, pressure and temperature stabilized, and forced through the column by an inert carrier gas or liquid mobile phase. Typical carrier or mobile phase gasses are nitrogen, helium, and argon. Carrier gasses are application specific and must not react with any component of the sample. Usable carrier gas is 99.99% pure and completely moisture free.

The separation section consists of various columns that segregate discrete components from the sample effluent. The sample is separated at this mobile phase interface by adsorption. Adsorption is the inclination of some substances to adhere to the surface of a particular media. This effect results in a stratification of compounds traversing a column of an adsorbent media. Compounds with no adsorbent reaction to the obstructive media travel through the column unimpeded. Compounds that are sensitive to the adsorbent media resist propagation and stratify along the column in degreed order of adsorption. Columns are made of small-diameter stainless steel tubing (1/8 inch or 1/4 inch) packed with different media to segregate particular substances, and wound into coils to conserve space, see Figure 6.4-2. Examples of stationary packing are porous silica microspheres, activated alumina, and activated carbon. Porous silica microspheres are often chemically coated with very selective compounds (substrates) to modify their characteristics. Continued introduction of additional reagent gas or solvent to the column forces the segregated components to

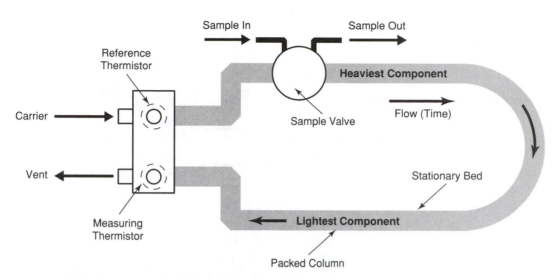

Figure 6.4-2 *Simplified Gas Chromographic Sample Column*

continue their migration through the packing. The sweeping of components from the column is called elution. The time from sample injection to the detection of a component is the elution time. Sample solution exiting the column contains the separated (eluted) compounds in relation to their propagation time. Elution time identifies the molecular size of the component. Components with the lightest molecular weight transverse the column first. Successive components eluted from the sample exit the column in order of increasing molecular weight. Elution time and component separation are a function of the selected packing and column length. If only lighter components are analyzed, a precut (or stripper) column is placed ahead of the analysis column to separate the lighter, faster components from the heavier, slower components. As soon as the light components exit the stripper column and enter the analysis column, column valves reverse flow to the stripper column and purge the heavy components from the column. Once purged of heavy components, the stripper column is placed back in service. By not having to wait on the heavy components to completely elute the entire column, sampling time is reduced and sampling rates are increased, thus resulting in increased efficiency.

6.5 Chromatograph Detectors

As the carrier/elute mixture exits the column, it is routed through the detection section. A detector measures each component in its separated state. The most common chromatograph detectors in use today are hydrogen flame ionization detectors, spectrometers, and thermal conductivity detectors. Flame ionization detectors are capable of detecting component concentrations as low as 0 – 1 ppm. As depicted in Figure 6.5-1, the detector consists of a chamber with inlets for hydrogen, column effluent (sample), and air, and an outlet vent. Inside the chamber are an air diffuser, a burner, an igniter, and the detection electrodes (anode/cathode). A DC voltage (100 volts – 1500 volts) is applied to the electrodes. The

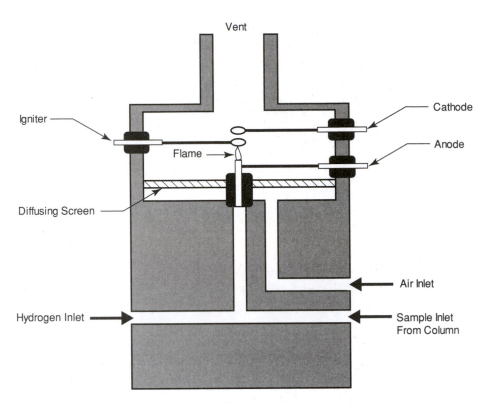

Figure 6.5-1 *Hydrogen Flame Ionization Detector*

hydrogen flame induces the emission of electrons from the hydrocarbon molecules in the effluent stream. The ions are collected on the electrodes and amplified in a high quality electrometer amplifier. The sensitivity of the detector is proportional to the carbon content of the mixture.

Infrared analyzers (spectrometers) are used to detect components that exhibit absorptive characteristics in the infrared region of the wavelength spectrum. When a molecule of gas (or liquid) is subjected to electromagnetic radiation it will selectively and disproportionably absorb a single specific wavelength of energy. Figure 6.5-2 compares the absorptive characteristics of several gases. This phenomenon permits qualitative analysis by spectral reactivity. The inverse of absorbance is transmittance. Transmittance is the fraction of radiant energy that, having entered a layer of absorbing matter, reaches its farther destination. Since the degree of transmittance is relative to sample concentration, this process also yields quantitative data. Plotting transmittance against wavelength results in a spectrogram. Spectral analysis is performed with a variety of wavelengths including ultraviolet, visible light, and infrared. Elemental gasses, hydrogen, nitrogen, oxygen, chlorine, and the rare gasses do not absorb energy in the infrared region. The spectrometer consists of a light source, a tunable grating or filter, a sample flow cell with detector, and a reference flow cell with detector, see Figure 6.5-3. The source emits a spectrum of radiant energy (light). Tunable gratings and filters isolate the precise wavelength absorbed by the product detected. This frequency is then transmitted simultaneously through the sample

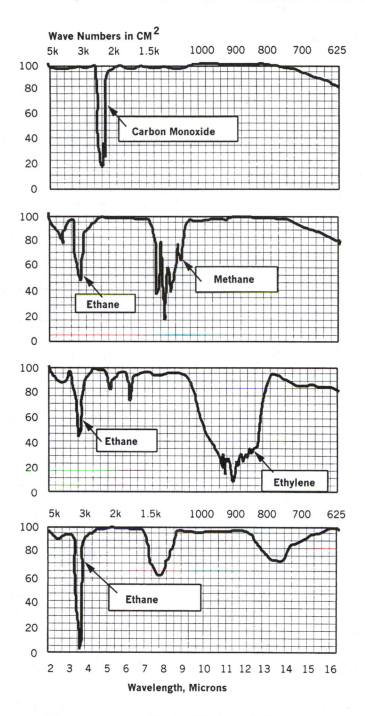

Figure 6.5-2 *Typical Infrared Spectra*

cell and the reference cell. The difference in exit energy detected between the sample cell and the reference cell is proportional to the energy absorbed by the sample. Since a linear relationship exists between absorption and concentration, quantitative analysis is permitted.

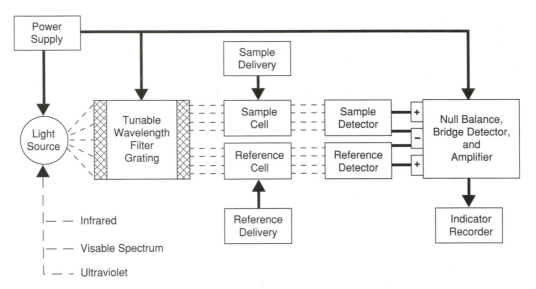

Figure 6.5-3 *Functional Block Diagram of Spectrometer*

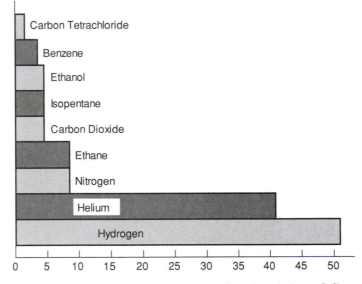

Figure 6.5-4 *Relative Thermal Conductivity of Gases*

Thermal conductivity analyzers are also used to detect components of chromatograph effluent streams. Thermal conductivity detectors consist of a Wheatstone bridge with a thermistor mounted in the measuring junction. The bridge is balanced to a null condition with pure carrier gas passing over the measuring thermistor. Sample component gases are less thermally conductivity than carrier gases. As the component gas passes over the measuring thermistor, the thermistors temperature rises decreasing the thermistor's resistance. The resulting bridge unbalance is representative of the sample concentration and elution time is representative of the component identity. Supporting amplifiers process the signal and the results are printed to a chart or bar graph.

Gas chromatograph malfunctions are usually related to sample delivery problems, sample contamination, or column packing contamination. Calibrations are performed by flowing special calibration gasses through the analyzer and adjusting for zero and span. Significant elements of chromatographic analysis are time, temperature, pressure, and flow. Practical applications strictly regulate sample temperature, pressure, and flow values. Sample propagation time, therefore, is the discriminating variable.

6.6 The Distillation Process

The distillation process adds heat energy to a liquid mixture forcing the compounds within the mixture to vaporize in order of volatility. The separated compounds are then condensed back into their liquid state. Volatile compounds readily transition to their gaseous state, while the heavier compounds tend to remain in the mixture. A popular distillation process increases the alcohol content of spirits by heating the water and alcohol mixture sufficiently to gas off the alcohol. The alcohol is then condensed (cooled) back into a liquid. Since the boiling points of water and alcohol are relatively close, the distillation process always collects some water and/or leaves behind some alcohol. Several repeat distillations are required to obtain significantly high alcohol concentrations. Simple distillation is generally sufficient to separate a single compound from a mixture.

Several different compounds are simultaneously extracted from a mixture by the fractional distillation process. Fractional distillation applies heat energy to a mixture just as simple distillation. Extracted vapors are then routed through a packed vertical column. Separated vapors stratify and congregate at specific levels in the column. Lighter compounds migrate to the top of the column while heavier compounds remain near the bottom. Extraction taps are located at different levels along the column. Each separated vapor is then drawn off and condensed into a pure product. Figure 6.6-1 depicts a simplified diagram of a fractional distillation column. Fractional distillation is an efficient process and is commonly used to refine petroleum. Gas chromatographic analysis can determine if a sample has components of ethane, ethylene, and propylene, and the concentration of each component. This trace detection ability is used to determine stratification of substances within a fractioning column. The amount of heat energy applied to the mixture determines the level of stratification within the fractioning column. If less heat is applied, the substances stratify at a lower level within the fractioning column. If more heat is applied, the substances stratify at a higher level within the fractioning column. Gas chromatograph sample lines are installed at each extraction tap. The chromatographs are configured to detect the predominate substance of the next lower or higher stratification level. A chromatograph configured to detect the substance at the next lower level of stratification expects to detect a minute trace of that product. If no trace is detected the temperature is too low, if a significant amount is detected the temperature is too high. A heat control signal is generated to correct temperature and bring the stratifications within the column back to the appropriate level.

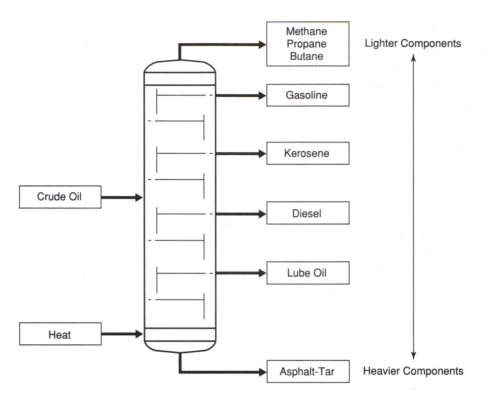

Figure 6.6-1 *Fractional Distillation Column*

▪ Chapter Six Summary

- Chemical analysis is the science and technology used to identify and measure the chemical composition of a product.
- Qualitative analysis determines the presence of a particular substance in a sample. Quantitative analysis determines the amount of a particular substance in a sample.
- Unstable compounds can exert tremendous energy if allowed to oxidize, and combustible processes can spontaneously ignite if exposed to oxygen.
- Measurement cells detect the concentration or activity of the sample, reference cells generate a constant fixed reference voltage, and temperature compensators provide a temperature feedback signal.
- Most reactions are temperature sensitive and require accurate temperature sensing feedback.
- Bridge circuits located in the analyzer detect the minute changes in cell signal values.
- pH analyzers measure the alkalinity or acidity of an aqueous solution.
- Highly acid concentrations are referenced at (0) on the pH scale. Pure water is neutral (7) and highly alkaline (base) concentrations are referenced at (14).
- Oxygen analyzer malfunctions include contamination and leaks.
- Conductivity analyzers detect the ability of a sample to support electrical current. Conductivity analyzers are commonly used to detect ionized particles in pure water.

- Conductivity (S) is the reciprocal of resistance. Therefore S = 1/ohm. The international unit of conductivity is siemens, often referred to as a mho.
- Specific conductance (K) is the conductance detected by one square centimeter probes through one cubic centimeter of sample.
- Precise (repeatable) values are best obtained with stable sample flow rates and sample temperatures.
- Representative values are obtained when the probe mounting location and sample line installation provide a thoroughly mixed and contaminant free sample solution.
- The most reliable test for a probe analyzer is to compare the readings to a known good sample (standard).
- A face mounted adjustment to correct the reading to match the standard solution value is called standardization.
- Temperature compensation resistance is measured with a digital multimeter in the ohms position. The ohm value obtained is referenced to an RTD response chart and compared to the known sample temperature. If the values agree, the temperature compensation probe is functional.
- To test the analyzer compensation circuits connect a precision resistance device adjusted for the correct ohm value to the analyzer temp comp connection and observe the analyzer reading. If the readings agree, the temp comp circuit is functional, and the probes are suspect.
- The spectrometer consists of a light source, a tunable grating or filter, a sample flow cell with detector, and a reference flow cell with detector.
- Adsorption is the inclination of some substances to adhere to the surface of a particular media.
- The stratification of compounds traversing a column of an adsorbent media results in a chromatogram.
- Chromatography is an analytical process that samples, separates, detects, qualifies, and quantitates a product.
- Multiposition sample valves extract a portion of the product and deliver the sample to the column.
- Column valves alternately route purge gas, sample solution, and sample gas through the column.
- Samples are forced through the column by an inert carrier gas or liquid mobile phase.
- A column of stationary phase packing separates the components in the sample.
- Sample solution exiting the column contains the separated compounds in relation to their propagation time.
- Gas chromatograph malfunctions are usually related to sample delivery problems, sample contamination, or column packing contamination.
- Significant elements of chromatographic analysis are time, temperature, pressure, and flow. Practical applications strictly regulate sample temperature, pressure, and flow values. Sample propagation time, therefore, is the discriminating variable.
- Chromatographs sample for trace elements of petroleum distillates at each level of a fractioning column and develop a temperature control demand signal to maintain proper stratification within the column.

Chapter Six Review Questions:

1. State the definition of chemical analysis.

2. State the purpose of qualitative analysis.

3. State the purpose of quantitative analysis.

4. State the purpose of analytical instruments.

5. State the undesirable effect of introducing oxygen into some processes.

6. State the purpose of a measurement cell.

7. State the purpose of a reference cell.

8. State the purpose of a temperature compensation detector.

9. Identify the acidity or alkalinity of a substance based on its pH value.

10. State two common oxygen analyzer malfunctions.

11. Describe the relationship between conductivity (S), resistance, and specific conductance (K).

12. State the most reliable method to determine operability of an analyzer probe.

13. State the adjustment used to correlate an analyzer reading to a known sample.

14. Describe the troubleshooting technique used to isolate a reference temperature malfunction.

15. Differentiate between simple distillation and fractional distillation.

16. Explain trace detection as used by chromatographs in control applications.

17. List three steps of the chromatographic process.

18. Describe the relationship between absorption and sample transmission time.

19. List several potential chromatograph malfunctions.

20. Calculate the ohm value required to simulate a conductivity reading of 2 milli-siemens from a cell constant 50 probe.

Chapter 7
Controllers

OBJECTIVES

Upon completion of Chapter Seven the student will be able to:

- *Describe the basic functions of pneumatic, electronic, and digital controllers.*
- *Describe automatic control mode, manual control mode, and cascade control mode.*
- *Develop the proportional response to both a step and ramped change in input.*
- *Develop the integral (reset) response to both a step and ramped change in input.*
- *State the effects of reset wind-up on controller output.*
- *Develop the derivative (rate) response to both a step and ramped change in input.*
- *State the purpose of bumpless transfer.*
- *List the functional components of a pneumatic controller.*
- *List the control and tuning adjustments available on a pneumatic controller.*
- *Describe a method to test operability and calibrate a pneumatic controller.*
- *Describe the function of pneumatic controller components including the input bellows, flapper nozzle, force balance beam, feedback bellows, and output relay.*
- *List the functional components of an electronic controller.*
- *List the control and tuning adjustments available on an electronic controller.*

- *Describe a method to test operability of and calibrate an electronic controller.*
- *Describe the function of electronic controller components including the voltage divider set-point, error detection amplifier, deviation circuit, integral circuit, and output driver.*
- *List the functional components of a digital controller.*
- *List the control and tuning adjustments available on a digital controller.*
- *Describe the function of digital controller components including the analog to digital converter, memory set-point, binary counter, algorithm processor, and digital to analog converter.*
- *List the common malfunctions associated with pneumatic controllers.*
- *List the common malfunctions associated with electronic controllers.*
- *List the common malfunctions associated with digital controllers.*

7.0 Introduction

Controllers provide process variable monitoring (input signal), set-point to process comparison (error signal), and developed correction signal algorithms (output). Transmitters and transducers provide the input signal to the controller. The controller receives and accepts the input as an accurate representation of the actual process variable monitored. The controller then compares the input signal to a predetermined optimal or desired value, the set-point. If a difference between set-point and input is detected, the controller computes and applies a specific preset algorithm output to eliminate the difference. The correction signal is transmitted to a final control element that manipulates the controlled variable to effect the correction. Controllers monitor inputs continuously and automatically respond to any changes. Specific correction factors required for each process vary. Several error response actions act individually, or combine, and provide the optimum correction signal. Scaled dials mounted directly on the controller provide response tuning and adjustment capability. Intelligent controllers store these values in memory and require tuning through a menu driven keypad or a communicator device. Position switches select automatic or manual control mode operation. Internal tracking of the output signal while in manual mode ensures a bumpless transfer when the controller is restored to automatic operation. Bumpless transfer is the ability to switch, or transfer, between automatic and manual control operation without disturbing the process or repositioning the final control element.

7.1 Controller Response

Controllers operate in several selectable modes. The most common are automatic, manual, and cascade. Automatic control is generally preferred. In this mode, the controller performs continuous monitoring and manipulation of the process as required to maintain set-point. Manual control overrides the automatic functions of the controller and allows the operator to use the controller output adjustment to position the final element in the desired position.

The automatic section of the controller tracks the manual output, and monitoring and recording functions are maintained. Cascade control is similar to automatic control, except the local set-point is disabled and another controller remotely controls the set-point.

Proportion band (1/gain × 100) P, integral (reset) I, and derivative (rate) D are the three most common response characteristics. Proportional values are expressed as percent of scale range and are generally between 0% and 500%. Proportional response characteristics are often expressed as gain. The reciprocals of these terms refer to identical response values. The terms are interchangeable providing correct conversions are applied. Proportion band is equal to 1/gain × 100 and gain is equal to 1/proportion band × 100. Proportion band determines response slope relationship between error, in percent of scale, and output, in percent of output range. Proportional correction is applied instantaneously and the response slope is fixed. The percent proportional band selected represents the amount of error, in percent of scale, required to produce a 100% change in controller output. A proportion band of 100% responds to a 100% error with a 100% change in output. A proportion band of 50% responds to a 50% error with a 100% change in output. A proportion band of 200% responds to a 200% error with a 100% change in output, Figure 7.1-1.

Integral (reset) is a time-weighted response. Since proportional response is instantaneous and fixed, proportional action alone may successfully stabilize the process at a value other than set-point. Integral action develops a correction value determined by the amount of error over time. Integral values are expressed as minutes per repeat or repeats per minute. The term "repeat" refers to the error signal. Integral action restores process to set-point by complimenting the original input signal value with a ramped amount of the error value and driving the output until the offset is eliminated, or the output saturates. A setting of one repeat per minute adds one error value to the output each minute, Figure 7.1-2. The error

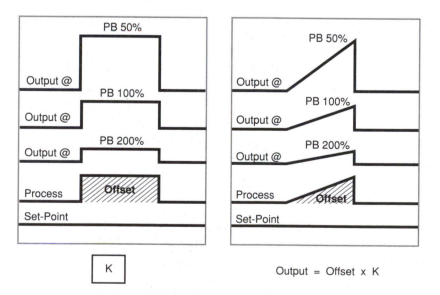

Figure 7.1-1 *Proportional Response*

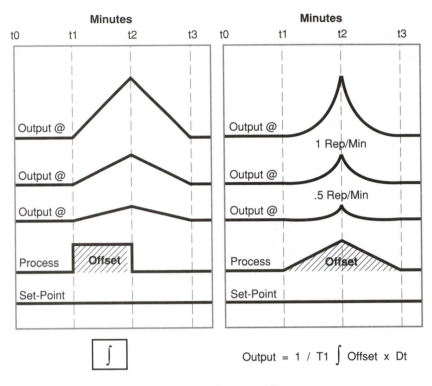

Figure 7.1-2 *Integral Response*

is reduced through each correction cycle (repeat) until offset is eliminated. Reset output continues until set-point deviation is eliminated. If the final control element fails to correct the offset, reset output saturates. This phenomenon is called reset wind-up. A controller in a reset wind-up condition does not respond to additional process changes until sufficient reset time integrals eliminate the wind-up.

Derivative (rate) is also a time-weighted response. Derivative action develops correction values relative to the rate of change of the input value, Figure 7.1-3. Derivative response is developed independently of error magnitude. This response characteristic is often referred to as lead. Proportional and integral are reactive correction characteristics, both actions only initiate in response to measured error. Derivative is a pre-active correction characteristic; response is initiated by a change of input value. Rapid input changes initiate an aggressive derivative response and moderate to gradual input changes initiate minimal to no derivative response.

The magnitude of each individual response characteristic is summed into a final cumulative value that constitutes the control algorithm, Figure 7.1-4. Controllers apply the algorithm to maintain process-input values at set-point. A correction signal is developed in response to deviation and the effectiveness of correction is determined by continually comparing process input (feedback) to set-point. Adjustments to the PID settings are referred to as tuning the controller.

146

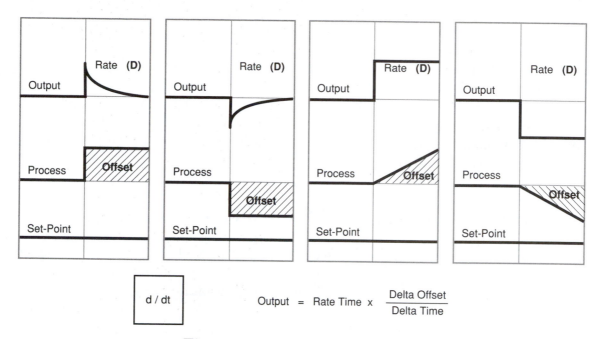

$$\text{Output} = \text{Rate Time} \times \frac{\text{Delta Offset}}{\text{Delta Time}}$$

Figure 7.1-3 *Derivative Response*

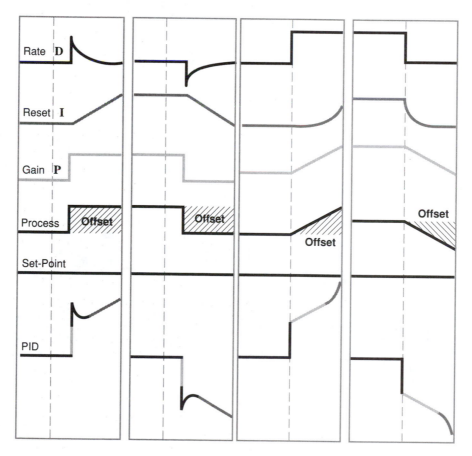

Figure 7.1-4 *Combined Response*

The controller effects all corrective action through a final control element. Ultimately loop stability and control is realized through final control element characteristics. Due to the variety of tuning parameters applied to different processes, control valve position is not always directly equivalent to controller output. The control valve positioner characterizes control valve position. The positioner accepts the demand (correction) signal from the controller and positions the valve according to a designed response profile. Common profiles are linear, quick opening, and equal percentage. These profiles are investigated further as they apply to final control elements.

7.2 Functional Components of Controllers

Pneumatic controllers monitor and compare a process variable to a set-point, detect the error, and apply an output algorithm to eliminate the error. Pneumatic controllers display the process variable value, set-point, and analog output value. Set-point adjustment and auto/manual transfer are also face mounted, Figure 7.2-1. Many controllers contain chart drive mechanisms to record process fluctuations. The chart drive is a clock-driven gear reduction roller that feeds scaled graph paper under the ink fed controller-indicating pens. PID tuning and chart drive on/off controls are located either on the rear or inside of the case. Partial removal of the controller from the panel is often required to access these controls.

The most common pneumatic signal is 3 – 15 psig. Pneumatic controllers incorporate a bellows assembly as the input sensor. The bellows converts input pressure into a physical displacement or motion to the balance beam, Figure 7.2-2. The beam is attached to the bellows at one end and a flapper/nozzle at the other. The beam pivots on a leverage

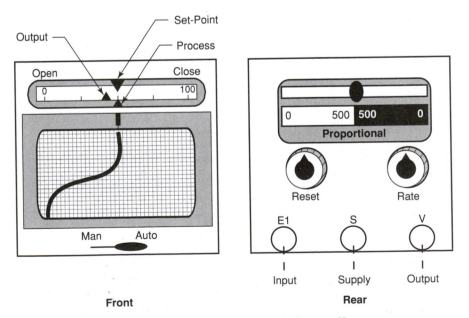

Figure 7.2-1 *Pneumatic Controller*

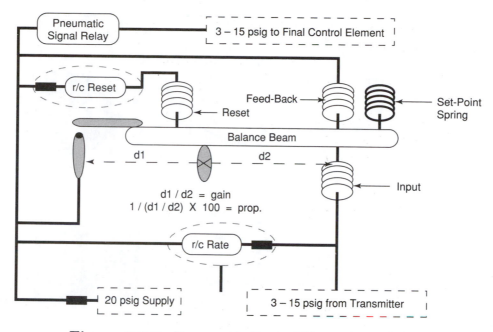

Figure 7.2-2 *Pneumatic Control Beam Assembly*

multiplier fulcrum. The flapper nozzle assembly detects induced displacement at one end of the beam as an error signal. The controller is supplied with 20 psig of continuous regulated instrument air (supply). The measured variable value from the transmitter is applied to the input bellows (input). The bellows transfers an equivalent amount of energy to a force balance mechanism. A flapper nozzle assembly detects the amount of deviation at the force balance beam. The flapper nozzle assembly backpressure is amplified by a pneumatic relay and retransmitted as a pneumatic output signal. Supply air flows through a restriction screw and is supplied in parallel to the nozzle, feedback bellows, and upper chamber of the output relay. The flapper acts to block the nozzle and build backpressure (error signal). The feedback bellows and output relay simultaneously sense the error signal. The feedback bellows expands to restore balance to the beam. The relay amplifies the error signal and outputs a representative value of supply air. Fulcrum position on the beam determines proportional band through simple lever ratios. A restrictor/tank R/C circuit, Figure 7.2-3, develops reset. The time constant of the tank network is a function of restrictor flow rate divided by tank capacity. Increasing or decreasing the amount of restriction adjusts reset action. The reset tank transmits to a reset bellows located on the beam to compliment the input bellows force. A tank circuit also develops the rate function. The rate tank however accepts input pressure and applies it directly to the output. Some rate tanks have an additional restrictor installed as a constant bleed.

Operability Assessment: Exercise caution when manipulating controller output. Final control devices may change state and introduce perturbations throughout the process. The primary function of any controller is to maintain process at set-point. If the controller output responds to correct a deviation between process and set-point, the controller is

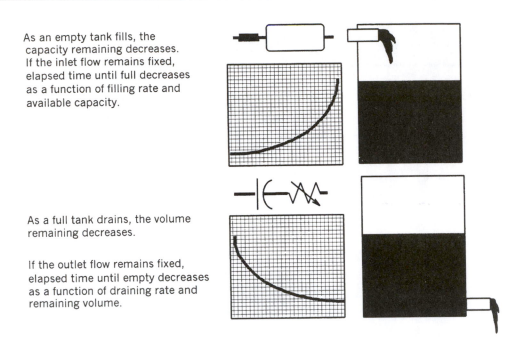

As an empty tank fills, the capacity remaining decreases. If the inlet flow remains fixed, elapsed time until full decreases as a function of filling rate and available capacity.

As a full tank drains, the volume remaining decreases.

If the outlet flow remains fixed, elapsed time until empty decreases as a function of draining rate and remaining volume.

Figure 7.2-3 *Resistive/Capacitive Timing Circuits*

functional. Although determined functional, calibration and/or tuning might increase the efficiency and stability of the process. Pneumatic controllers are susceptible to many of the common pneumatic woes. The most common pneumatic controller failures are related to inadequate or unclean supply air. Air supply to pneumatic instruments is routed through a filter regulator. Routine maintenance includes verifying proper regulator adjustment and periodically blowing down the filter trap at the petcock. Pneumatic relays are provided with a restriction cleanout plunger or removable restrictor. Cleanout plungers are affixed to the relay so that pressing the plunger drives a small wire through the fixed restriction and removes any accumulated debris. Removable restrictors are removed and cleaned periodically. Misalignment and wear of the flapper nozzle assembly is also a common failure; perform the manual flapper/nozzle test. Check for air leaks at the relay mounting block and feedback bellows.

Calibration: Indicator section: Apply lower range value to controller input and adjust indicator as required to obtain 0% indication. Apply upper range value to controller input and adjust input transducer linkage to correct one-half of observed error at 100% indication. Repeat sequence until all values are within desired tolerance. If a linearity error exists after offset and span are calibrated, apply midrange value to controller input and adjust drive linkage as required to move indicator five times the amount of error in the direction of the error. Offset and span may require additional adjustment.

Controller Section: Place set-point at 50%. Apply a measured midrange input value to the controller. Place proportional band to 100% and/or gain to 1. Place reset to minimum minutes per repeat or maximum repeats per minute. Place derivative to minimum/off.

Adjust flapper/nozzle to obtain 9 psig measured at the controller output. Allow sufficient time for the reset bellows to obtain 9 psig, approximately one minute. Place reset to Off to trap air in reset bellows. This ensures the beam is balanced during calibration. Apply a measured lower range input value to the controller, and adjust set-point indicator as required to observe 3 psig at output. Apply a measured upper range input value to the controller, and adjust set-point linkage as required to obtain 15 psig at output. If upper and lower range stops are installed, ensure output drops to zero at bottom stop, and output ramps to full value at upper stop.

Reset: Ensure proportional band is placed at 100% and reset is set at minimum time. Ensure derivative function is "off." With controller in automatic place set-point to 50% and apply input of midrange value. Allow controller output to stabilize at 9 psig. Place reset to "one" and introduce a 25% set-point deviation. The output should ramp 3 psig in one minute. Derivative: Ensure proportional band is placed at 100% and reset is set at minimum time. Ensure derivative function is "on." With controller in automatic place set-point to 50% and apply input of midrange value. Allow controller output to stabilize at 9 psig. Place reset to "off" and introduce a 25% set-point deviation. The output should step and then decay. See Figure 7.2-4.

Electronic controllers perform the same basic functions as pneumatic controllers. Although executed by electronic circuits, the control functions of proportional, integral, and derivative are identical. The analog process variable display output level indicator, set-point indicator,

Figure 7.2-4 *Level Controller*

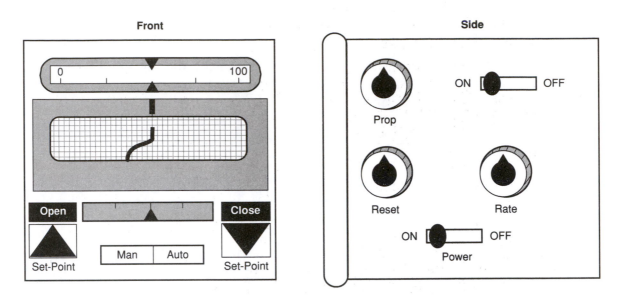

Figure 7.2-5 *Electronic Controller*

set-point adjustment, and auto-manual transfer switch is located on the front of the controller. The chart drive data logger mechanism employs a printer cartridge or a thermal writing device. A mechanical latch located on the front of the controller secures the controller in the panel. Once released, the controller is partially pulled from the panel to access the tuning knobs located inside the case, Figure 7.2-5.

A differential operational amplifier circuit, Figure 7.2-6, generates proportional band. A voltage divider circuit on the reference input establishes set-point, and a feedback resistor develops gain (proportion). An electrical R/C tank composed of a resistor and a capacitor develops integral. A R/C tank also develops derivative.

Calibration: Electronic controllers are generally field calibrated by injecting a monitored input value, typically 4 – 20 mADC, 1 – 5 DC volts, or 0 – 10 DC volts. Offset and span pots are accessed through the top of the case. In some cases, cable driven indicator arrows require centering on the cable. This is done by inputting a mid-scale value, loosening the clamp, and gently sliding the arrow to indicate midscale. Some electronic controllers are supplied with a sensitivity adjustment. Increase sensitivity until the arrow vibrates, and then decrease sensitivity until stability returns. Monitor controller current outputs by placing a DMM in ampere mode in series with the output wiring. Monitor controller voltage outputs by piggybacking the output terminals with a DMM in the Volts mode.

Digital controllers perform the same basic function as pneumatic and electronic controllers; they do, however, have several added advantages. Digital controllers offer the ability to configure an array of functions. Digital controllers are also capable of communicating in multidrop configurations, and performing auto tuning and self-diagnostics. Analog process value, set-point, and output value are displayed on an LCD screen, Figure 7.2-7. Data trends are stored in memory or written to a floppy disk. Tuning parameters are accessed through

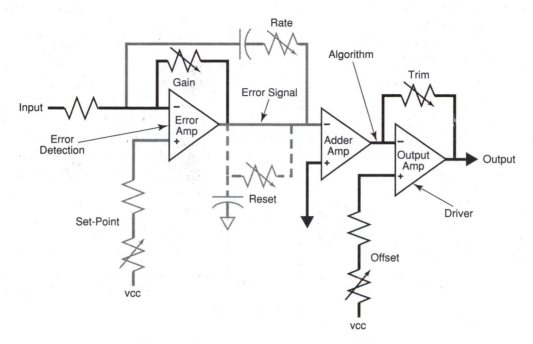

Figure 7.2-6 *Operational Amplifiers*

a menu-driven keypad or a digital communicator. Signal processing is accomplished by several standard digital components. The analog-to-digital converter accepts the process variable and converts the analog value into an equivalent binary count. The count is then

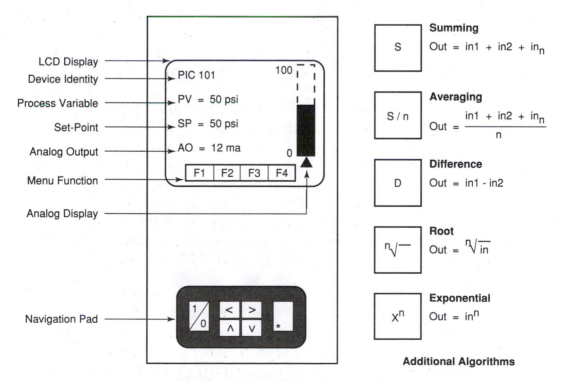

Figure 7.2-7 *Digital Controller*

153

transmitted to a binary counter/comparator that determines the difference between the process value and the memory set-point stored in a register. The algorithm processor uses the extracted error value to develop the appropriate correction value. The final digital correction signal is then converted back into an analog value by the digital-to-analog converter and transmitted to the final control element by an isolated output driver.

Calibration: To calibrate a digital controller, access the calibration prompt through the menu driven keypad. The menu leads through a calibration sequence that requests the LRV and URV. Apply these values when requested, but do not select until the values are stable. An output calibration is also often required. Enter the calibration software through the keypad, and monitor the output values and trim as required. It is important to "send," "save," or "write" any changes made to ensure default values are not restored after a power interrupt.

7.3 Local Controllers

Local controllers perform identical functions to board mount controllers; they are, however, normally enclosed in weather-tight housings and mounted in the field near the controlled device. Local controllers provide local indication of process variable value, set-point adjustment and indication, output indication, manual output adjustment, and adjustment dials for proportional, integral, and derivative. Many local controllers provide output balance indicators to assist bumpless transfer functions, Figures 7.3-1. Recording controllers trend data on a circular chart.

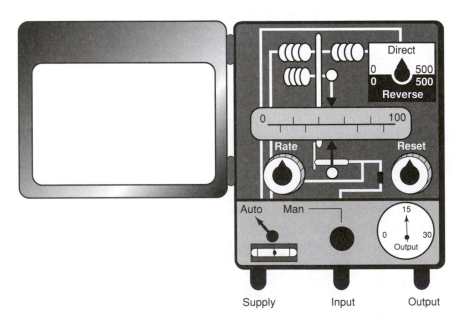

Figure 7.3-1 *Local Controller Adjustments*

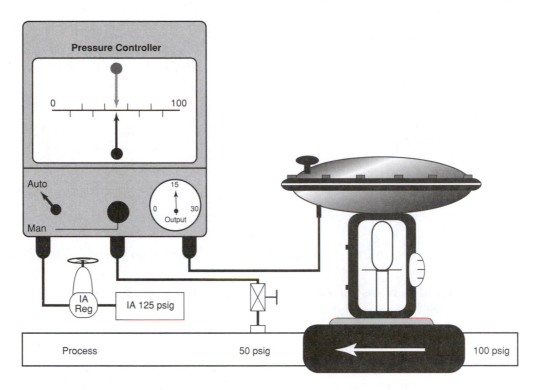

Figure 7.3-2 *Local Controller with Valve*

Local controllers are often physically mounted, as shown in Figure 7.3-2, to the final control element (air operated valve). The controller input is connected to the process line and the process variable is detected directly at the controller. The controller output is transmitted directly to the valve actuator. The valve is adjusted to position from full open to full closed in relation to the local controller output. The need for a transmitter in the control loop is eliminated. Construction and maintenance costs are reduced, but control is not as responsive or precise as conventional transmitter, controller, and positioner control schemes. See Figure 7.3-3.

Operability Assessment: As automatic controllers are designed to compensate for deviations in the control loop, they are seldom responsible for hard failure of the control loop. Provided an accurate input signal and an operable final control element, controllers automatically adjust demand until process is at set-point.

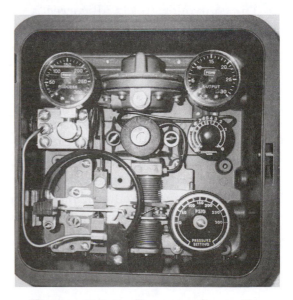

Figure 7.3-3 *Pressure Controller*

Electronic controllers are extremely dependable devices. The most common failures are associated with the pen and paper drive mechanisms. Most control loop failures are introduced by ancillary devices or cabling problems. Digital controllers rarely fail and control extremely well if programmed properly. Almost all digital controllers have an on-board diagnostic program to verify the electronics; most problems, however, occur with the LCD display or keypad.

■ Chapter Seven Summary

- The controller compares the input signal to a predetermined optimal or desired value, the set-point.
- The controller computes and applies a specific preset algorithm designed to offset or eliminate the error.
- The correction signal is transmitted to a final control element that physically manipulates the controlled variable to effect the correction.
- Bumpless transfer is the ability to switch between automatic and manual control operation without disturbing the process or repositioning the final control element.
- Proportional values are expressed as percent of scale range and are generally between 0% and 500%. Proportion band is equal to 1/gain × 100 and gain is equal to 1/proportion band × 100. Proportion band determines response slope relationship between error, in percent of scale, and output, in percent of output range.
- Integral (reset) is a time-weighted response. Integral action develops a correction value determined by the amount of error over time. Integral values are expressed as minutes per repeat or repeats per minute.
- Derivative action is a pre-active correction characteristic that develops correction values relative to the rate of change of the input value.
- The controller effects all corrective action through a final control element.
- The most common pneumatic signal is 3 – 15 psig. Pneumatic controllers incorporate a bellows assembly as the input sensor. The flapper nozzle assembly detects bellows induced displacement at one end of the beam as an error signal. The flapper nozzle assembly backpressure is amplified by a pneumatic relay and retransmitted as a pneumatic output signal.
- Fulcrum position on the beam determines proportional band through simple lever ratios.
- Electronic controllers perform the same basic functions as pneumatic controllers. Although executed by electronic circuits, the control functions of proportional, integral, and derivative are identical.
- A voltage divider circuit on the reference input establishes set-point and a feedback resistor on the differential operational amplifier circuit determines gain (proportion).
- An electrical R/C tank composed of a resistor and a capacitor develops integral. An R/C tank also develops derivative.

- Digital controllers offer the ability to configure an array of functions. Digital controllers are also capable of communicating in multidrop configurations, and performing auto tuning and self-diagnostics. Tuning parameters are accessed through a menu-driven keypad or a digital communicator.
- Many local controllers provide output balance indicators to assist bumpless transfer functions.
- The most common pneumatic controller failures are related to inadequate, wet, or unclean supply air.
- The most common electronic controller failures are associated with the pen and paper drive mechanisms.
- Digital controllers rarely fail and control extremely well if programmed properly. Some problems do occasionally occur with the display or keypad.

Chapter Seven Review Questions:

1. Describe the basic functions of pneumatic, electronic, and digital controllers.

2. Develop the proportional response to a step change in input.

3. Develop the proportional response to a ramp change in input.

4. Develop the integral (reset) response to a step change in input.

5. Develop the integral (reset) response to a ramp change in input.

6. Develop the derivative (rate) response to a step change in input.

7. Develop the Derivative (Rate) response to a ramp change in input.

8. State the purpose of bumpless transfer.

9. List the functional components of a pneumatic controller.

10. State the location of the control and tuning adjustments available on a pneumatic controller.

11. Describe the function of pneumatic controller components including the input bellows, flapper nozzle, force balance beam, feedback bellows, and output relay.

12. List the functional components of an electronic controller.

13. State the location of the control and tuning adjustments available on an electronic controller.

14. Describe the function of electronic controller components including the voltage divider set-point, error detection amplifier, deviation circuit, integral circuit, and output driver.

15. List the functional components of a digital controller.

16. State the location of the control and tuning adjustments available on a digital controller.

17. Describe the function of digital controller components including the analog-to-digital converter, memory set-point, binary counter, algorithm processor, and digital-to-analog converter.

18. List the common malfunctions associated with pneumatic controllers.

19. List the common malfunctions associated with electronic controllers.

20. List the common malfunctions associated with digital controllers.

Chapter 8

Final Control Elements

OBJECTIVES

Upon completion of Chapter Eight the student will be able to:

- *Describe the different control valve types, including spring diaphragm, dual acting piston, motor operated valves, hydraulic operated valves, and solenoid valves.*
- *Describe the purpose and function of the basic components of an air operated control valve including the I/P transducer, positioner, actuator, valve limit switches, actuator mounted solenoids, booster, and valve body.*
- *Differentiate between balanced and unbalanced trim.*
- *Describe the ANSI seal off categories.*
- *List several hazards and state the safety precautions associated with AOV calibrations.*
- *Describe the proper calibration process for a positioner.*
- *Describe the proper stroke length adjustment process.*
- *Describe the proper bench-set adjustment process.*
- *Interpret a diagnostic scan to determine the operability characteristics of an AOV including bench-set, spring rate, seat load, stroke length, air set, and friction.*
- *Define the function of the basic components of a motor operated control valve including the drive motor, torque limiter, rotary variable displacement transducer, and travel limit switches.*
- *Define the function of the basic components of a hydraulic operated control valve including the servo card, servo valve, and linear variable displacement transformer.*

- *State the purpose of the hydraulic pump, hydraulic reservoir, and accumulator.*
- *Describe the technique to calibrate a servo driver card.*
- *List and explain the function of the basic components of a solenoid operated control valve including the coil, rectifier, and plunger.*
- *State the preferred method to determine valve operability.*
- *List malfunctions causing the failure of a valve to open/close.*
- *List malfunctions causing a failure to control (track demand signal).*
- *List malfunctions causing oscillation, hunting, or sticking.*
- *List the effects of control valve malfunction on process parameters.*

8.0 Introduction

Several types of final control elements are available. Control valve types include spring diaphragm, dual acting piston, motor operated valves, hydraulic operated valves, and solenoid valves. The operational and control demands of the specific application determine selection. Final control devices are identified by their actuator design and by their body design, Figures 8.1-1 and 8.1-2. The source of motive force employed to position the valve classifies the type of final control element. Air operated valves (AOVs) convert air pressure into motive force. They are assembled with either spring diaphragm actuators or dual acting piston actuators. AOVs are used in many applications and are quite common. AOV advantages include rapid response and excellent control characteristics. Since typical plant air supply pressures are limited to less than 150 psig, maximum force capabilities are limited by air pressure values. Motor operated valves (MOVs) and solenoid operated valves

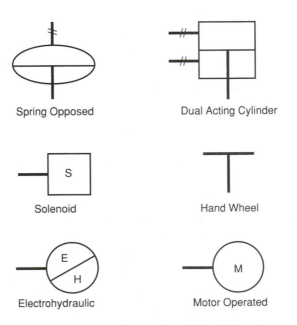

Spring Opposed

Dual Acting Cylinder

Solenoid

Hand Wheel

Electrohydraulic

Motor Operated

Figure 8.1-1 *Actuation Identification*

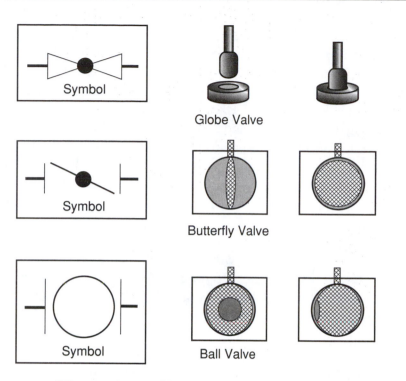

Figure 8.1-2 *Valve Body Identification*

(SOVs) convert electromotive force into valve motion. The MOV actuating device is an electric motor. MOVs generate substantial force but do not have the rapid response capability of air operated valves. MOVs are capable of modulated control, but are more commonly used in on/off applications. The SOV actuating device is a coil. SOVs are relatively inexpensive, but energized coils of reasonable size generate minimal force. Hydraulic operated valves convert hydraulic fluid pressure into valve motion. As typical hydraulic pressures can reach 2000 psig, hydraulic actuators generate extreme amounts of force. Hydraulic valves provide rapid and accurate response. The disadvantages of hydraulic actuators are related to the expense and complexity of the pumps, reservoirs, accumulators, and support devices required.

8.1 Fundamental Components of Final Control Devices

Regardless of the actuator type, all valves except solenoid valves have similar body and bonnet assemblies. The bonnet attaches the actuator to the valve body. A packing or stuffing box located in the bonnet provides a sealing zone that allows stem motion but prevents process leakage. Several types of packing are available and selection is based on process reactivity, temperature, and pressure. Teflon chevron packing produces the least resistance to valve stem motion, but use is limited to low temperature applications. Braided asbestos packing is used in moderate to high temperatures. Graphite impregnated tape packing is used in extreme temperature applications. Packing does degrade over time and periodic adjustment is required. Packing adjustment nuts are installed on the valve bonnet assembly to facilitate packing adjustment. Self-adjusting constant load packing is available. The traditional packing nut washers are replaced with a stack of dished (Bellville) washers and torqued to a set value; as the packing consolidates over time the dished washers expand to maintain constant packing torque. Excessive packing friction can affect overall valve performance, including insufficient seat seal off, stem binding, erratic positioning, and reduced stroke time. Loose or worn packing results in process leakage. The valve body encapsulates the valve trim and attaches the valve assembly to the process piping. It provides and maintains the process pressure to environmental sealing boundary at the process connections. Several process piping attachment techniques are employed including flanged, threaded, and welded.

Valve trim includes all internal process wetted parts but is specifically the plug, stem, cage, seat, and guides. Common designs include plug in seat "globe valve," piston in cage "caged globe valve," Saunders "boot valve," butterfly, and ball. Plug, cage, and Saunders trims are sliding stem designs, while butterfly and ball valves are rotary designs. Trim sets are physically configured to provide specific flow curves. Linear trim produces a direct linear relationship between valve position and flow. Quick opening trim produces higher flow rates earlier in the valve stroke. Equal percentage trim produces an exponential flow rate in relation to valve position. Butterfly and ball valve trim sets rotate 60 to 90 degrees and

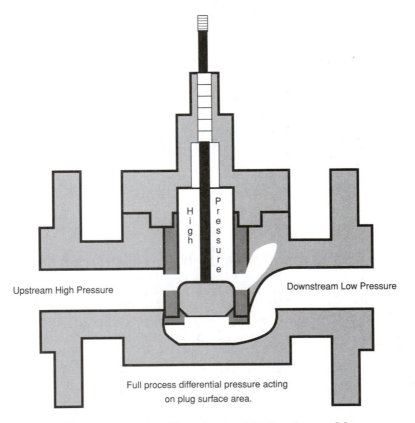

Upstream High Pressure   Downstream Low Pressure

High Pressure

Full process differential pressure acting
on plug surface area.

Figure 8.1-3 *Unbalanced Trim Assembly*

approximate quick opening flow characteristics. Globe type trim sets are either balanced or unbalanced. Saunders valves are often ceramic lined for use in corrosive service. The operation is similar to globe type valves, except instead of a seat and plug, the Saunders valve actuator attaches to an isolating diaphragm/plug assembly that eliminates the need for packing. The valve closes to restrict flow by pressing the soft plug assembly into a mated contour in the lower valve body. These valves are generally used for corrosive liquids or sludge. Figures 8.1-3 and 8.1-4 demonstrate the differences.

The simplest concept is unbalanced; this means that when the valve is closed all upstream pressure is acting on the plug. The unbalanced plug in the closed position is subjected to the complete differential pressure that exists across the valve. If the flow direction is down through the valve, the full differential pressure across the valve is assisting to keep it closed. If the flow direction is up through the valve, the pressure differential is working to open the valve. The balanced trim set has balancing ports drilled through the plug. The balancing ports allow the valve outlet pressure to act equally on both sides of the plug. Unbalanced valves generate higher seat load/seal off pressures and are used where positive seating is a priority. Large, unbalanced valves often require a small parallel bypass line or an internal pilot plug to reduce differential across the valve and assist them open. Control valve trim

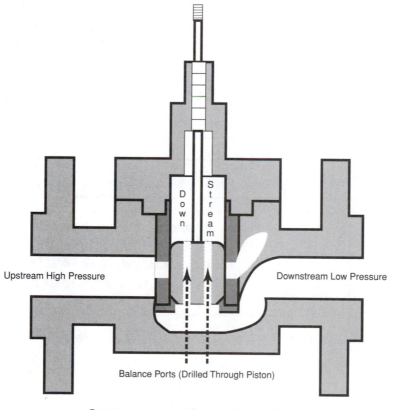

Downstream pressure acting on both top and bottom of plug.
Differential pressure equalized.

Figure 8.1-4 *Balanced Trim Assembly*

ANSI Leak Class	Maximum Allowable Leakage	Test Medium	Test Pressure	Test Procedure Requirements
I	N/A	N/A	N/A	None
II	0.5% rated capacity	air or water	45-60 psig	valve inlet pressurized valve outlet vented
III	0.1% rated capacity	air or water	45-60 psig	valve inlet pressurized valve outlet vented
IV	0.01% rated capacity	air or water	45-60 psig	valve inlet pressurized valve outlet vented
V	0.0005 ml per minute water per inch of port diameter per psid	water	maximum process D/P >/= 100 psig	valve inlet pressurized valve outlet vented valve body filled with water and stroked close
VI	one bubble per minute for a one inch port three bubbles per minute for a for a two inch port 27 bubbles per minute for a six inch port	air or nitrogen	maximum process D/P not > 50 psig	valve inlet pressurized valve outlet vented normal actuator thrust stabilization time allowed

Ansi Leak Class	Minimum Pounds Per Lineal Inch of Port Circumference at 1000 psid.	
II	50	*Values are approximations to assist understanding of the above table.*
III	100	
IV	150	*Obtain actual values per ANSI document B16.104 current revision.*
V	250	

Figure 8.1-5 *ANSI Leak Class Table*

flow capacity is specified as Cv. The Cv value is the amount of flow, in gallons of water per minute, at standard pressure and temperature, which produces a differential pressure drop of one psig across the wide-open valve. Rangeability is the ratio of maximum controllable valve flow to minimum controllable valve flow. Valve sizing requirements dictate that a control valve trim set large enough to control extremely large flow rates is naturally design

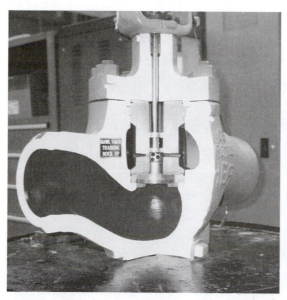

Figure 8.1-6 *Globe Valve Trim (Cutaway)*

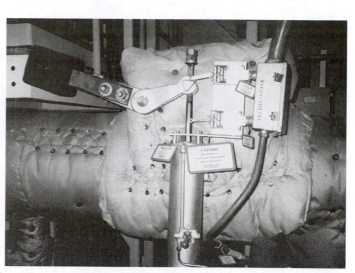

Figure 8.1-7 *Travel Limit Position Switches*

limited at controlling minute flow rates. One solution to systems that demand a high degree of control over an extremely wide range of flow rates is to install a smaller valve in parallel with a larger valve. This configuration is similar to the bypass valve installation used to reduce differential across large unbalanced trim sets. In either case, both valve positioners are driven from the same I/P or controller. A split-range calibration is established at the valve positioners that strokes the small valve full range with the first fifty percent of controller output and strokes the large valve full stroke with the final fifty percent of controller output. The calibration values are typically 3 − 9 psig = 0 − 1 inch travel for the smaller valve, and 9 − 15 psig = 0 − 4 inch stroke for the large valve. These values control the smaller valve at lower demand for good low flow controllability. As the small valve reaches the upper limit of its flow capacity, the large valve begins to open. At high flow rates, the small valve remains full open and the large valve provides good controllability at the higher flow rates. Valve trims are also classified in accordance to seal off capability.

ANSI leak class definitions specify the maximum allowable leakage for valves in each category. The designations range from Class I, which requires no testing, to Class VI, which permits an almost imperceptible amount of leakage. Figure 8.1-5 describes the leak rates and testing requirements for each ANSI leak classification. Valve seal off performance is actually determined by the amount of closing force available per lineal inch of seat. Valve body design, flow direction, and the amount of unbalanced seat area all contribute to this value. Valve position is indicated by a pointer on the stem that reads out on a scale attached to the valve body. Accessories include limit switches, solenoids, and boosters. Limit switches provide remote indication of valve position and valve open/close logic to program controllers. Solenoids enable and inhibit valve actuation and provide failsafe control of the valve by logic controllers. Boosters multiply air volume from the positioner signal to reduce valve stroke times. See Figures 8.1-6, 8.1-7, 8.1-8, and 8-1.9.

Figure 8.1-8 *Control Valve Positioner*

Figure 8.1-9 *Air Supply Regulator and Positioner Feedback Linkage*

8.2 Operation of Air Actuated Valves

Air operated valves are available with spring diaphragm or dual acting piston actuators. The basic components of an air operated control valve are the actuator, bonnet, body, and trim. Auto/manual controllers generate an analog position demand signal to the air operated valve. On/off position demand to air operated valves is determined by open/close pushbuttons or logic controller actuated solenoids. Limit switches affixed to the valve energize open/close position lights. Many AOVs are supplied with a hand wheel to permit manual operation. Spring diaphragm actuators are assembled with a flexible diaphragm assembly opposing an internal spring. Air pressure acts on the diaphragm to overcome the spring tension and position the valve. When air pressure is removed, the internal spring returns the valve to its fail-safe position. Valve travel is limited by diaphragm convolution flexibility, and closing force is limited by diaphragm pressure operating range. Dual acting actuators are assembled as cylinders with an internal piston plate. Air supplied to the top of the piston forces the valve downward, and air supplied to the bottom of the piston forces the valve upward. The valve maintains a fixed position when upper cylinder and lower cylinder air values are equal. With no spring installed, fail-safe position is indeterminate. Typical aluminum cylinders operate at higher-pressure values than diaphragms, and therefore generate additional closing force. Piston actuators also offer increased stroke

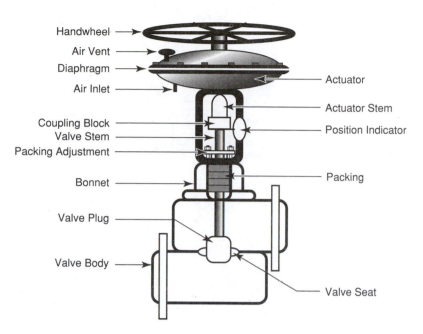

Figure 8.2-1 *Air Operated Valve*

length. The functional components of an air operated control valve include the actuator, bonnet, body, and trim, and may include an I/P transducer and valve positioner. Accessories include limit switches, solenoids, and boosters. The I/P transducer converts the controller current demand signal into a proportional pneumatic signal and transmits it to the control

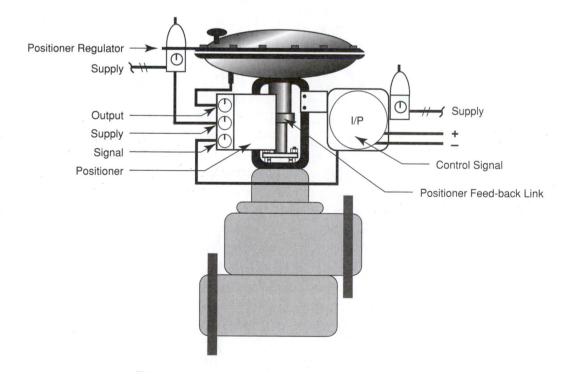

Figure 8.2-2 *Valve with Positioner and I/P*

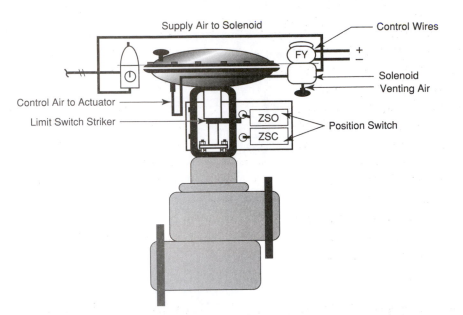

Figure 8.2-3 *Valve with Solenoid and Switches*

Figure 8.2-4 *Air Supply Valve with Positioner and Solenoid*

valve, or positioner if provided. The positioner compares the air signal from the I/P to the valve position and generates an open or close signal to obtain the desired valve position. The positioner-input diaphragm accepts the air signal from the I/P, and the positioner balance beam monitors actual valve position through a feedback link and cam assembly. The force balance mechanism detects any difference between position demand and actual valve position. The output relay drives a pneumatic signal to the actuator that repositions the valve, restores the beam to a balance condition, and eliminates any difference between demand position and actual position. The actuator receives and geometrically multiplies the positioner pressure signal. The actuator develops force as the positioner output pressure is multiplied by the net effective area of the actuator and is converted into linear motion (thrust). The actuator then transmits the thrust/motion to the valve stem through a mechanical coupling. The bonnet assembly attaches the actuator to the valve body. See Figures 8.2-1, 8.2-2, and 8.2-3.

Operability Assessment: Failure to open or close is usually due to a loss of demand signal (I/P), loss of supply air (regulator), actuator failure, or trim failure. Failure to control (track demand signal) is usually caused by water in the airline, insufficient air supply pressure or volume, I/P out of calibration, positioner out of calibration, or valve specifications inappropriate for the application. Excessive positioner or booster gain; a worn, loose, or binding feedback link; or a loose positioner-mounting bracket induces oscillation. Hunting is often caused by the positioner input bellows leaking signal air, water in the instrument airline, insufficient supply pressure or volume, positioner gain too low, or the actuator diaphragm leaking by. Sticking is the result of excessive packing friction or internally damaged components. The most conclusive technique to determine valve operability is to stroke the valve full range with the controller and compare controller output signal to positioner input signal and valve position at several points. Verify that positioner input is proportional to controller output.

8.3 Alignment and Calibration of Air Operated Valves

Alignment: These values are established at initial valve calibration and rarely, if ever, change. Initial control valve alignment is normally only required if the valve was disassembled for rework.

 Control valves open and close with tremendous force and can easily remove a human appendage. Valve position during this process is always determined by applying a regulated air supply directly to the actuator.

Do not attempt to control valve position during these adjustments with the valve positioner. The actuator may change position abruptly in an uncontrolled manner and cause serious personal injury.

Never exceed the manufacturer's recommended pressure rating. Actuator area and spring rate are both fixed values determined in the selection process. Apply a regulated air supply to the actuator and drive the actuator to the minimum and maximum recommended pressure values. Measure and record the distance traveled. Ensure that the actuator travels smoothly in both directions, that no air leaks are present, and that actuator stroke exceeds the specified valve stroke.

Regulate air pressure to retract the actuator rod and carefully insert a spacer between the actuator rod and the valve stem. Slowly increase/decrease the pressure to the maximum/minimum rated value and verify the stem is driven to the seat. Retract the actuator rod and carefully remove the spacer from between the actuator rod and the valve stem. Turn the spring adjustment nut until the actuator stem just begins to tension at the desired upper bench set value. This is sometimes evident as a slight movement or as a rapid increase in actuator pressure in response to a steady ramped input. This adjustment fixes the upper bench set value. Observe the actuator stem and do not allow it to turn when adjusting the spring range nut. If the actuator rod is allowed to turn, it will induce a twist into the diaphragm and lead to premature failure of the actuator.

Apply sufficient pressure to the actuator to extend the stem the desired stroke length and attach the coupling block, mating the actuator to the valve stem. The amount of air pressure required to offset spring tension with the actuator rod extended is determined by the spring rate, effective actuator area, and stroke length. This is the lower bench set value and represents the amount of closing force available. Adjust the scale to indicate Valve Closed.

Apply sufficient pressure to place valve at midposition. Attach the valve positioner feedback arm, and verify the arm is at a ninety degree angle and the positioner cam is at fifty percent travel. If the valve has limit switches mount them equal distance from the striker plate. Slowly stroke the valve full travel in both directions and verify sufficient positioner clearance and cam travel are available. Adjust limit switches as required. Proceed to calibration.

CALIBRATION:

I/P: I/P transducer calibration is covered in a previous chapter.

Regulator: Adjust regulator to valve/positioner maximum operating value. The preferred technique is to monitor the regulator output while it is attached to the positioner. This ensures that the regulator is not deadheaded.

Positioner: To calibrate the positioner apply the lower range value to the positioner input, typically 3.25 psig, and verify or adjust positioner zero adjust until positioner output just begins to ramp. Apply the upper range value, typically 14.75 psig, then verify or adjust positioner range adjust until the positioner output saturates to full supply pressure.

Positioner output should rapidly drop to zero at any input value less than 3.25 psig. Positioner output should rapidly ramp to full airset value at any input value greater than 14.75 psig.

Stroke length: Stroke length is the distance, in inches, the valve stem travels from stop to stop. Valve travel is referenced to zero when the plug is firmly into the seat. The distance in inches the stem retracts before contacting the actuator stop is the stroke length. Available excess actuator travel is indicative of seating force. Stroke length is field adjustable. However, a freshly lapped plug and seat establishes optimum surface mating shortly after the valve is placed in operation, and rotating the valve stem may invalidate previous blue check verifications. Measure or scribe the current "valve closed" position. Apply sufficient air to place actuator at midstroke. Slightly loosen stem lock bolts.

Attempting to turn stem with valve closed will damage the seating surface.
Do not loosen stem lock bolts with valve at either end of travel.
Stem thread damage can occur.

Count number of threads per inch on valve stem. A stem thread pitch of eight threads per inch will change stem extension length 1/8 inch per stem revolution. Screw the stem "in" to increase valve stroke length, and screw the stem "out" to decrease valve stroke length. Adjusting the stem out to decrease stroke length increases seating force and raises the lower bench set value. Adjusting the stem in to obtain increased stroke length reduces seating force and decreases the lower bench set value. If the stem does not have wrench flats, use a strap wrench to prevent scoring the stem and resultant packing leakage. If turning the stem out does not reduce stroke length, the valve was probably not originally seated.

8.4 Air Operated Valve Diagnostic Analysis

Several valve manufacture and service companies have developed diagnostic testing equipment for air operated valves. Intelligent positioners even carry limited diagnostic software on-board. The most significant advantage of diagnostic equipment is the ability to capture and compare real-time diagnostic response data. Keypad sequences differ from system to system, but the same operational principles apply. The system consists off a laptop computer that houses a signal generator, loads the software, and provides an operator interface, a set of pressure transducers, a strain gauge, and the associated tubing, cabling, and clamps required to install the transducers. Pressure monitoring transducers are usually tapped into the airset regulator, positioner input, and actuator input. A travel transducer and strain gauge are attached to the valve stem. The signal generator is wired to the I/P.

Several test sequences are programmed into the testing software. Once the test sequence is initiated, the diagnostic software ramps or steps the current to the I/P and records thousands of data points as the valve responds. Typical test programs include a dynamic ramp scan, static points scan, and a step study scan. The dynamic scan ramps a 4 – 20 mADC input to the I/P and monitors all points. Dynamic scan data, Figure 8.4-1, reveals seat load, bench set, spring rate, friction, regulator airset, dynamic positioner and I/P calibration and response, and stroke length. A skilled analysis of scan data also reveals seat profile characteristics, air leaks, and internal damage. A static point scan, Figure 8.4-2, is similar to a traditional five-point calibration, however further analysis reveals oscillation and overshoot not readily apparent during a traditional static calibration. A step study scan inputs an incremental series of increasing and decreasing steps. This scan is often used to locate or isolate problems occurring within specific areas of valve operation. The equipment is menu driven and only moderately difficult to operate; the full benefits, however, are only realized when experienced operators review the scan sheets.

Much of the dynamic principles and terminology are generally helpful in understanding the operation of an air operated valve. The following exercise incorporates a simplified technique to obtain dynamic values with traditional testing equipment.

Actuator Effective Area: Pressure is equal to force applied over an area. Actuators act like force multipliers. A typical actuator may have one hundred square inches of effective

171

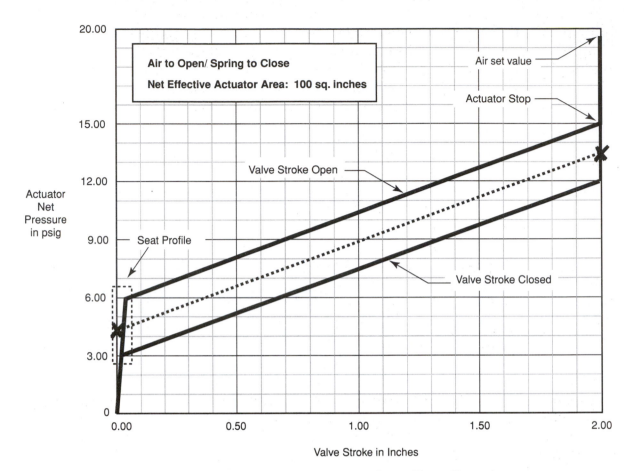

Figure 8.4-1 *Dynamic Valve Scan Data Sheet*

area. This means that if a pressure of 15 psig is applied to the actuator it would develop a force of 15 psig × 100" square = 1500 pounds. This is sufficient force to compress a 500-pound per inch spring three inches. Actuator area and spring size are matched at the factory and are not altered without serious consideration and calculation. Manufacturer's publications specify effective area for each actuator. To validate actual effective area, compare measured actuator air pressure to actuator stem force as measured with a strain gauge. Strain gauge indicated force divided by measured actuator air pressure yields true effective area.

Spring Rate: Spring rate is determined at the point of manufacture by such parameters as coil diameter, coil wire core diameter, and tensile characteristics of the coil wire. These values change very slowly under normal conditions and the springs seldom require replacement. Spring rate values are specified as pounds per inch. This value remains constant over the entire usable range of compression. A variable rate spring will exhibit multiple compression values as it compresses, but they are seldom used in air operated valves. A spring designated as a 450-pound-per-inch spring requires 450 pounds of force to compress one inch. This same spring requires 900 pounds of force to compress two inches, and only 225 pounds of force to compress one-half inch.

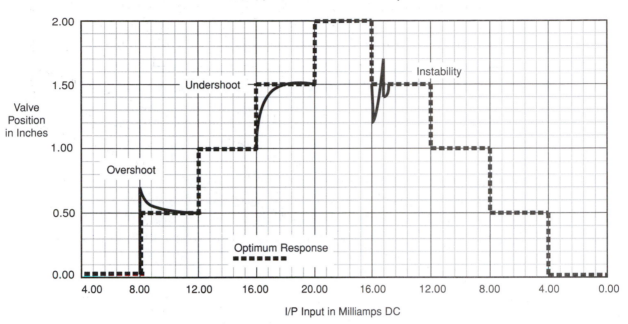

Figure 8.4-2 *Static Point Scan*

Friction: All valves should exhibit some friction. The amount of friction expected is dependent on the type of valve packing used and the valve stem diameter. Teflon packing produces minimal friction, while live load packing friction values on a large valve are substantial. To determine friction compare the actuator air pressure value required to position the valve at midstroke from both increasing and decreasing directions. The difference in pressure values multiplied by the net effective actuator diaphragm area is equivalent to the frictional resistance of the valve. Since friction resists valve motion equally in either direction, divide the value obtained by two. Excessive friction or a significant deviation in friction over the valve stroke indicates binding or internal damage.

Example: An actuator with net effective area of 100 square inches requires 10.5 psig to position mid stroke from the closed position, and only 7.5 psig to reach the same position from an open direction. The difference of these two values is 10.5 − 7.5 = 3.0 psig.

Dividing by two results in 3.0/2 = 1.5 psig

Multiply by area to determine force 1.5 psig × 100" square = 150 pounds

Friction equals 150 pounds. This is a low number, indicative of Teflon.

Seat Load: Seat load is the amount of excess closing force available after the valve stem has contacted the seat area. If we assume a lower bench set value of 4.5 psig, this means the actuator requires 4.5 psig to offset spring compression at that point. If our actuator had an effective area of one hundred square inches, that would allow 4.5 psig × 100 square inches = 450 pounds excess closing force, or seat load.

Bench Set: *Determined by dynamic parameters:* The bench set value is representative of the force required to offset spring tension. The classic terminology defines bench set as the values obtained excluding all frictional forces. Bench set is accurately determined through deduction of dynamic parameters. Verify the valve begins to stroke at the positioner output value equivalent to the valve's lower bench set value, then verify the valve travel is full stroke at the positioner output value equivalent to the valve's upper bench set value. Record these values and obtain the same readings from the opposite direction. You should have four numbers: (1) actuator air pressure as valve just begins to open, (2) actuator pressure as valve just reaches full open, (3) actuator pressure as valve just begins to go closed, (4) actuator pressure as valve just reaches full closed. The hysteresis evident in these readings is induced by internal valve friction. Since normal frictional forces are equal and opposite, they are self-canceling. Actual bench set values are determined by taking the difference between the two lower readings and the difference between the two upper readings. Divide the differences by two, and add them back to the smaller of the original values. For the purpose of the example, assume 100 square inch effective area, 2-inch stroke length, and 450 pounds per inch spring.

Example:
(1) Actuator pressure as valve just begins to open: 6 psig
(2) Actuator pressure as valve just reaches full open: 15 psig
(3) Actuator pressure as valve just begins in closed direction: 12 psig
(4) Actuator pressure as valve just reaches closed position: 3 psig

Difference between (1) and (4) =   6 psig - 3 psig = 3 psig
Difference between (2) and (3) = 15 psig - 12 psig = 3 psig

Divide differences by two (friction is applied in both directions)
    3 psig/ 2 = 1.5 psig   3 psig/ 2 = 1.5 psig

Add back to lesser values, upper and lower:
Bench set lower = 4.5 psig   Bench set upper = 13.5 psig

    Friction = 1.5 psig × 100 square inches = 150 pounds
    Seat load = 4.5 psig × 100 square inches = 450 pounds
Stroke force = upper bench set (13.5) minus lower bench set (4.5) = 9.0 psig

9.0 psig × effective area = 9.0 psig × 100 square inches = 900 pounds
Stroke force/spring rate = 900 pounds / 450 pounds per inch = 2 inches

Valve stroke = 2 inches

8.5 Motor Operated Valves

Motor operated valve (MOV) position is initiated by operator station pushbuttons on a control panel or by logic controller switches. An internal gear reduction assembly converts rotational motor torque into linear valve motion. Closing force is limited by current capacity of the motor control. Internal travel limit switches that energize open/close position indication lamps determine valve position indication, and travel limit switches that enable or interrupt motor voltage implement valve travel control. Motor controlled valves are usually supplied with hand wheels to permit valve operation in the event of power failure or motor malfunction. An engagement clutch is used to select manual or motor operation. The functional components of a motor operated control valve are the drive motor, gear reducer, torque limiter, rotary variable displacement transducer (RVDT), travel limit switches, position indication switches, and valve body. The drive motor is a high torque electric motor. The gear reducer assembly converts motor rotation into linear valve stem travel. The torque limiter is a clutched disengagement assembly designed and adjusted to protect the valve trims from damage. The RVDT is a precision rotation detector; it is used to provide feedback for analog operation of rotary travel valves. Travel limit switches interrupt voltage to the motor at preset valve travel limits to prevent over torque and excessive current demand. Position indication switches energize and illuminate open and closed position lights and provide valve position logic to program controllers.

8.6 Solenoid Valves

The functional components of a solenoid operated control valve include the coil, rectifier, plunger, and valve body. The coil provides motive force to position the plunger. The rectifier converts AC voltage into DC voltage. The plunger changes position to redirect process or air flow through the solenoid. Solenoid activation is initiated through operator-controlled pushbuttons, pressure, flow, or temperature switches, or program logic controllers. Some solenoid valves have a manual operator such as an external plunger or rotary shaft that permits manual operation of the valve. Manual operators inhibit normal operation of the solenoid if they are left in the engaged position. Solenoid valves are commonly used to select on/off control of air operated valves, collect samples, and provide trip circuits on compressors and turbines. The common tubing is attached to the actuator and solenoid actuation selects between supply air and atmospheric vent, or between normal flow and trip flow. The common port is usually identified as (1), (A), or cylinder. The other ports are labeled as (2) or (B), and (3) or (C).

 The operation of many solenoid valves is determined by the tubing configuration. The control action can be changed from normally closed to normally open by reconfiguring the attachment tubing. Prior to disconnecting a solenoid, label and record the tubing and port assignments. Ensure each tube is reconnected to the appropriate port.

Solenoid Enclosures:		
Type 1 Standard	Type 4 Watertight	Type 7 Explosion Proof

Coil Operating Voltages:			
A.C. Voltage		**D.C. Voltage**	
Nominal Voltage	Voltage Range	Nominal Voltage	Voltage Range
24	20 - 24	6	5.1 - 6.3
120	102 - 120	12	10.2 - 12.6
240	204 - 240	24	20.0 - 25.0
480	408 - 480	120	102.0 - 126.0
		240	204.0 - 252.0

Coil Class Temperature Limits: (Celsius)			
A	B	F	H
105	130	155	180

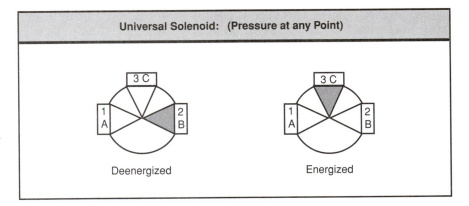

Universal Solenoid: (Pressure at any Point)

Deenergized   Energized

Figure 8.6-1 *Typical Solenoid Data*

Ensure the solenoid design matches the installation requirements. Housing enclosures are available in different environmental ratings including weather tight and explosion proof. Solenoid valve body ports are available with different orifice sizes. Port orifice size can effect system response time, valve stroke time, and sample collection volume. Solenoid valves are available in both DC and AC configurations of various voltage ratings and are not interchangeable. Typical DC voltage ratings are 6, 12, 24, 120, and 240. Typical AC voltages include 24, 120, 240, and 480. The coils are also temperature rated. Figure 8.6-1 displays typical rating information. Always refer to the specific manufacturer's specifications for the details of each device.

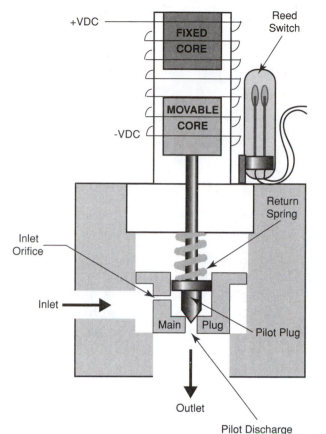

Deenergized Closed Position:
Inlet orifice bleeds sufficient inlet pressure into chamber above main plug to achieve full rated delta pressure across seat for positive seal off.

Energized Open Position:
Solenoid coil exerts sufficient energy to lift pilot plug. Pilot discharge port dumps faster than smaller inlet orifice can replenish. Chamber above main plug drains/vents down to equal outlet pressure. Higher inlet pressure lifts main plug.

Reed Switch Assembly:
Reeds must be perpendicular to magnetic core for maximum sensitivity. As movable core pulls in to open position, the proximity of the magnetic core pulls the switch contacts together and completes switch circuit.

Operability:
Valve may not operate properly without sufficient process pressure or flow.

Figure 8.6-2 *Pilot Operated Solenoid Valve*

Operability Assessment: Operational coils are hot to the touch when energized. Degraded coils often emit a buzzing sound. The vent port should not weep when de-selected. Ensure the vent screen or muffler is not obstructed.

High pressure, high flow, and certain extreme service solenoid requirements exceed the force capabilities of common coils. These solenoids use a coil to operate a small internal pilot and depend on process differential to stroke the main plug. Figure 8.6-2 demonstrates pilot valve operation.

8.7 Hydraulic Operated Valves

Hydraulic operated valves are common in applications requiring precise control of large flow rates. Hydraulic valve position demand is generated through a servo-control unit or panel-mounted operator push buttons. Linear variable displacement transformers (LVDTs) provide position feedback, and limit switches illuminate open/close lights. Typical installations are turbine driven pumps and generators. Functional components of

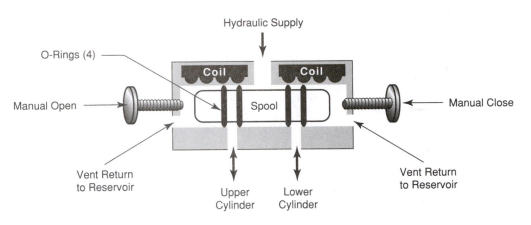

Figure 8.7-1 *Hydraulic Shuttle Valve (Valve Stable)*

a hydraulic operated control valve include the actuator, servo valve, servo card, LVDT, hydraulic supply system, and valve body. Control valve action is initiated through pushbuttons or speed control devices. Limit switches are often installed to provide position indication and logic. The actuator is a heavy-duty steel cylinder with an internal piston. Hydraulic fluid pressure applied above the piston provides downward force, and fluid pressure applied below the piston provides upward force. Most hydraulic actuators have extremely large springs that provide stability and fail-safe assist. The servo valve is attached between the hydraulic fluid supply and drain lines and the actuator. The servo valve alternately directs hydraulic fluid supply pressure to one side of the piston while

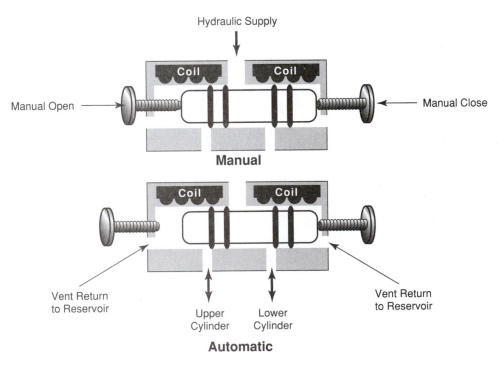

Figure 8.7-2 *Hydraulic Shuttle Valve (Valve Opening)*

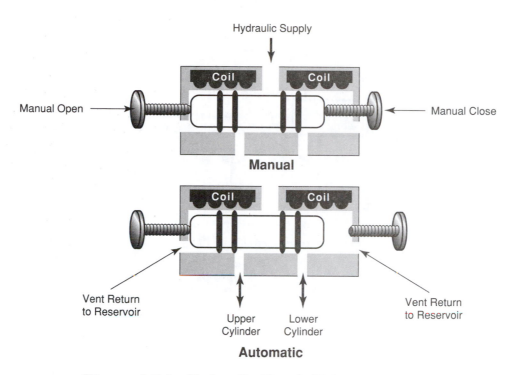

Figure 8.7-3 *Hydraulic Shuttle Valve (Valve Closing)*

draining fluid from the other side. The servo valve can also hold fixed pressure across the piston to stop the valve at midposition. Figures 8.7-1, 8.7-2, and 8.7-3 show three conditions of operation. Servo valve bias is adjusted to fail the hydraulic valve to the fail-safe position. The servo card is usually mounted in a remote controller unit. It compares position demand to actual valve position and transmits a correction signal to the servo valve. The LVDT linear variable differential transformer is attached to the hydraulic cylinder and valve stem. It converts valve position into a voltage signal and transmits that information to the servo card.

Calibration: Servo controller cards are calibrated similar to electronic controllers. Optimum results are obtained if the LVDT is centered. Drive the valve from full closed to full open and record the LVDT feedback voltage at both points. Ensure these values are equal and opposite. To adjust the LVDT, drive the valve to midposition and adjust the LVDT mounting rod to obtain zero volts. Apply a monitored valve-closed demand signal to the servo card, typically 4 mADC. Adjust the offset potentiometer until the valve just contacts the seat. Apply a full-open demand to the servo card, typically 20 mADC. Adjust gain until the valve just reaches full open. Repeat the process until all parameters are satisfactory. Some servo cards provide a dither adjustment. Dither is a small-magnitude AC signal component added to the servo output. The purpose of dither is to maintain a slight state of excitation at the valve at all times to promote longevity and reduce the buildup of varnish and sludge within the servo valve. Servo valves often have a bias adjustment located on the valve body. This adjustment determines the fail position of the hydraulic actuator. Fail-safe

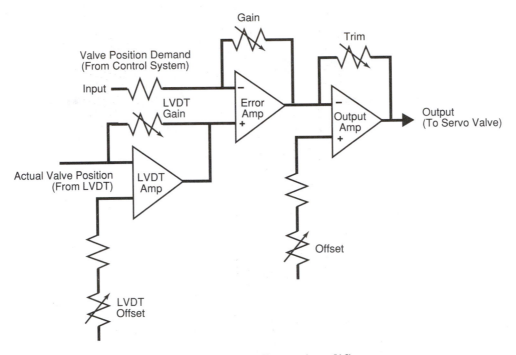

Figure 8.7-4 *Servo Amplifiers*

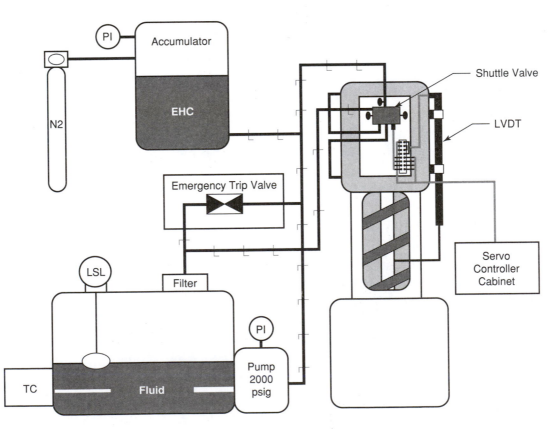

Figure 8.7-5 *Electro-Hydraulic Valve Support System*

position of a servo valve is determined by measuring servo card output voltage with the valve stable at midposition. Zero volts indicate a fail in position servo. A fail-closed valve requires some negative offset to hold midposition, nominally (-) 200 millivolts. A fail-open valve requires some positive offset to hold midposition, nominally (+) 200 millivolts.

The hydraulic supply system in Figure 8.7-5 consists of a high-pressure pump, a reservoir, an accumulator, and various heaters, coolers, and filters. The hydraulic pump supplies high-pressure fluid to the valve. The hydraulic reservoir provides a safe, clean, controlled supply of fluid to the pump. The accumulator provides an inert gas blanketed, remote reservoir of hydraulic fluid to the system. The accumulator assists in maintaining regulated pressures throughout the system. It increases overall system response and reduces cycling of the pump.

Figure 8.7-6 *Electro-Hydraulic Control Oil Skid*

Proper maintenance of electrohydraulic control systems includes periodic filter replacement, reservoir level verification, reservoir vent desiccant replacement, accumulator blanket charging, visual leak inspection, and hydraulic fluid chemistry monitoring.

■ Chapter Eight Summary

- The basic components of an air operated control valve are the actuator, bonnet, body, and trim.
- The I/P transducer converts controller current demand into a proportional pneumatic signal and retransmits it to the positioner.
- The positioner accepts the position demand signal from the I/P and drives a pneumatic signal to the actuator that repositions the valve.
- The actuator receives the positioner output pressure, converts it into linear motion (thrust), and then transmits the thrust/motion to the valve stem through a mechanical coupling.
- Valve stem packing in the bonnet provides a sealed orifice that allows stem motion but prevents process leakage.

- Excessive packing friction results in insufficient seat seal off, stem binding, erratic positioning, and reduced stroke time. Loose or worn packing results in process leakage.
- Limit switches provide remote indication of valve position and valve open/close logic to program controllers.
- Solenoids enable and inhibit valve actuation and provide fail-safe control of the valve by logic controllers.
- The functional components of a motor operated control valve are the drive motor, gear reducer, torque limiter, rotary variable displacement transducer (RVDT), travel limit switches, position indication switches, and valve body.
- The drive motor is a high-torque electric motor that provides force to the gear reducer assembly where it is converted into valve stem travel.
- The torque limiter is a clutched disengagement assembly designed and adjusted to protect the valve trim, gear reducer, and motor from damage.
- The RVDT provides position feedback for analog operation.
- Travel limit switches interrupt voltage to the motor at preset valve travels limits to prevent over torque and excessive current demand.
- Functional components of a hydraulic operated control valve include the actuator, servo valve, servo card, linear variable displacement transformer (LVDT), hydraulic supply system, and valve body.
- The servo card compares position demand to actual valve position and transmits a correction signal to the servo valve.
- The servo valve alternately directs hydraulic fluid supply pressure to one side of the piston while draining fluid from the other side to position the valve, or holds a fixed pressure across the piston to stop the valve at midposition.
- The accumulator assists in maintaining regulated pressures throughout the system. It increases overall system response and reduces cycling of the pump.
- Proper maintenance of electrohydraulic control systems includes periodic filter replacement, reservoir level verification, reservoir vent desiccant replacement, accumulator blanket charging, visual leak inspection, and hydraulic fluid chemistry monitoring.
- The functional components of a solenoid operated control valve include the coil, rectifier, plunger, and valve body.
- Initial valve calibration establishes the point that the valve begins to stroke, the valve's lower bench-set value, and the point that the valve travel is at full stroke, the valve's upper bench-set value.
- Excessive positioner or booster gain, a worn, loose, or binding feedback link, or a loose positioner-mounting bracket induces oscillation.
- Hunting is often caused by the positioner input bellows leaking signal air, water in the instrument airline, insufficient supply pressure or volume, positioner gain too low, or the actuator diaphragm leaking by.
- Unfavorable results of control valve malfunction include poor quality, low yield, equipment damage, environmental contamination, personal injury, or loss of human life.

Chapter Eight Review Questions:

1. Describe the different control valve types including spring diaphragm, dual acting piston, motor operated valves, hydraulic operated valves, and solenoid valves.

2. State the purpose of an I/P transducer.

3. State the purpose of a positioner.

4. State the purpose of an actuator.

5. State the purpose of valve stems packing.

6. State the location of valve packing adjustments.

7. State the effects of improper packing adjustment.

8. State three types of valve body trim assemblies.

9. State the purpose of valve limit switches.

10. State the purpose of actuator-mounted solenoids.

11. State the purpose of boosters.

12. State the purpose of the basic components in a motor operated control valve including the drive motor, torque limiter, RVDT, and travel limit switches.

13. Explain the function of the basic components of a hydraulic operated control valve including the servo card, servo valve, and LVDT.

14. State the purpose and explain the function of the hydraulic pump, the hydraulic reservoir, and the accumulator.

15. List the basic components of a solenoid operated control valve.

16. State the preferred method to determine valve operability.

17. Identify valve performance parameter established at initial valve calibration.

18. List malfunctions causing a failure to open/close malfunction.

19. List malfunctions causing a failure to control (track demand signal).

20. List malfunctions causing an oscillation, hunting, or sticking.

Chapter 9

Emergency Shutdown and Interlock Systems

OBJECTIVES

Upon completion of Chapter Nine the student will be able to:

- *State the purpose and intent of emergency shutdown systems (ESDs).*
- *Describe the use of layered safety.*
- *List the fail-safe modes of a valve.*
- *State the purpose of fail-safe valves.*
- *List five fail-safe support devices.*
- *Define the term single point failures.*
- *Define the term mitigate.*
- *State the fundamental requirement of a fail-safe system.*
- *Describe several methods used by ESD systems for securing the process.*
- *Define and explain redundancy.*
- *Define and explain the advantages of diversity.*
- *Define and explain the advantages of separation.*
- *State the benefits of redundant controls.*
- *State the drawbacks of redundant controls.*
- *State three methods used to determine if redundant control is required.*
- *State the purpose of interlocks and permissives.*
- *State the purpose of trips.*

- *List six process variables typically monitored by interlocks, trips, or permissives.*
- *List the machine variables typically monitored as interlocks, trips, or permissives.*
- *List the motive medias used to effect interlocks, trips, and permissives.*

9.0 Introduction

The purpose and intent of ESD (emergency shutdown systems) is to eliminate unnecessary trips or transients due to single point failures, reduce or mitigate undesirable effects of human performance errors, provide a safe and controlled response to unexpected conditions, protect equipment from damage, protect the environment from contamination, and prevent personal injury. These systems are also called safety instrumented systems, safety shutdown systems, and safety interlock systems. There are many-engineered safety features designed into critical systems to secure the process. The extent of safeguards incorporated into a particular system is determined by the inherent volatility of the process and the negative impact potential resulting from a loss of control. A desired safety integrity level is determined based on engineering review of design based impact assessments, industry events, and plant operating experience. Safety integrity levels range from one to three with one being the lowest level of safety performance and three the highest.

Emergency shutdown systems are designed with several layers of protection. The obvious layers are the design of the instrument control system that actually initiates the response. Other layers include security, operating and testing procedures, training and certification, work control guidelines, environmental controls, and labeling. Securities enhancements restrict physical access to equipment and controls, require passwords to access software, and require write protect features on SMART field devices. Operating procedures clearly define the appropriate response to any safety system status information or annunciation. Testing procedures control the extent and frequency of equipment operability and calibration testing. Testing procedures often include time response testing. These tests use real time data loggers and recorders to determine the total safety system response time delay from the initial triggering event to the full application of automatic safeguards. Many systems that use solid-state relay logic incorporate low voltage test circuits to validate continuity of the switches and relays. A "tickle" voltage, 12 – 18 volts, is applied to the actual actuation relays. Since the minimum pick-up voltage of the relays is much greater than the test voltage, the relays do not actuate. The test voltage does, however, illuminate low voltage current limited test lamps that are wired with the relay. Training and certification programs ensure that only qualified individuals operate or maintain the system. Work control guidelines ensure that all work activity is in accordance with the safety design documentation, and prevent unauthorized changes. Environmental controls are sometimes required to comply with manufacturer's device operability requirements. The warranted life of many components is calculated at relatively mild ambient conditions. It is sometimes more economical to provide environmental controls than perform the additional testing required to extrapolate the combined effect of an adverse environment. Adverse environmental effects include temperature, humidity, electromagnetic/radio frequency interference, shock, vibration, electrostatic discharge, radiation, contamination, and dust.

9.1 Redundancy

Installation of redundant instrumentation is an expensive but extremely effective safeguard solution. There are two redundancy architectures, identical redundancy and diverse redundancy. Identical redundancy provides several identical channels to simultaneously monitor the same process parameter. In Figure 9.1-1, each channel monitors the process as well as the other companion channels. If any single channel generates a significantly different reading, it is flagged as a deviant channel. The suspect channel is then automatically removed from service and placed in by-pass. The process remains unaffected and continues to operate normally on the remaining channels. As an added benefit, single channels are removed from service to support on-line maintenance and testing without affecting overall system safety or response.

Diverse redundant architecture provides several types of devices or technologies to execute equivalent safe shutdown sequences. Diversity is applied at all layers of protection. Some identical redundant architecture uses multiple channels of similar devices, but the devices are supplied by alternate manufacturers. Separation criteria are often developed to prevent a single adverse condition from disabling all redundant systems. Safety system wiring is segregated from process control wiring. Redundant safety system wiring is often routed

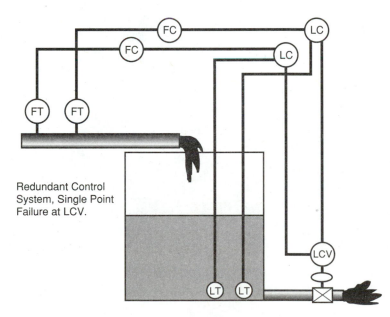

Figure 9.1-1 *Redundant Safeguard*

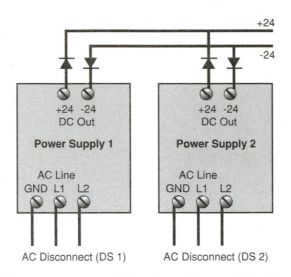

Figure 9.1-2 *Redundant Power Supplies*

through separate cabling trays. If one cabling tray is damaged by fire, flood, or heavy equipment, safety system wiring in the other tray remains intact and operable.

Redundant channels should be clearly marked and readily identifiable. Color schemes are often used to distinguish separate safety channels. The color scheme often extends to include control panel dials and switches, and field cabling. Redundant power supplies are common. A dual power supply system only loads a single power supply unit, Figure 9.1-2. If one power supply fails in service, the other instantly accepts load. The unloaded power supply is replaced on-line without interruption. Sourcing AC line power from different disconnects

Figure 9.1-3 *Redundant Trip Solenoids*

abates single point failure due to switchgear failure or operator error. Figure 9.1-3 shows dual redundant trip solenoids powered from separate sources and connected to the valve with stainless steel flexible hose.

9.2 Fail-safe Design

Properly designed fail-safe systems are not dependent on external energy sources to default the configuration of final control elements to a safe position. A loss or interruption of electrical, hydraulic, or pneumatic control power initially transfers critical components to a redundant power source. If the back-up power source fails to actuate, critical components default to their shelf (de-energized) state and initiate a safe shutdown sequence. Control valves are configured for specific fail-safe positions and particular failure modes. Failure mode selections include fail-as-is, fail-open, or fail-closed. Visual inspection can often determine the fail mode. With the exception of valves with reversed trim sets, fail-open valves require air to close on top of the diaphragm, fail-close valves require air to open on the bottom of the diaphragm, and fail-as-is valves accept air to the top and bottom of the diaphragm. Fail modes are designed to configure or restore the process to the safest or most desirable condition in the event of a loss or degradation of control system capabilities. Additional fail-safe support devices include solenoids, springs, accumulators, and shuttle valves.

9.3 Interlocks, Trips, and Permissives

The purpose of interlocks, trips, and permissives is to automatically interrupt or reconfigure final control device(s) if monitored variables significantly deviate from specifications. They allow equipment to start and operate only when monitored variables are within designed parameters, Figure 9.3-1. Typical process variables monitored are process flow, level, temperature, pressure, chemistry, contamination, and reactivity. Typical machine variables monitored are coolant level, temperature, and pressure; lubricant level, temperature, and pressure; and machine vibration, temperature, and speed. Trips automatically de-energize or turn off equipment if monitored variables exceed preset values, Figure 9.3-2. Interlocks and permissives inhibit unanticipated actuation of equipment and ensure correct start-up/shutdown sequences are followed. Interlock designs are mechanical, hydraulic, pneumatic, or electrical. Electrical interlocks are either hardwired (relay logic) or software (processor controlled).

 Monitoring trip and/or permissive contacts with a DMM in the ohm or ampere position will initiate the trip and/or permissive sequence.

Figure 9.3-3 depicts a mechanical overspeed trip device. This device is common on large rotating equipment. The trip plunger is recessed into a cylindrical bore perpendicular to the

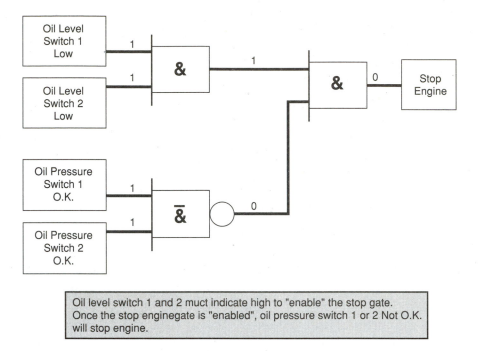

Figure 9.3-1 *Logic Trip String*

shaft. A counterspring is installed to a specific compression to resist the plunger inertia at normal operating speeds. If machine speed exceeds the desired limit, the plunger generates sufficient inertia to compress the spring. As the spring compresses, the plunger protrudes from the bore and actuates the trip lever. As in Figure 9.3-4, the mechanical overspeed and manual trip pushbutton often trigger the same trip lever. This particular arrangement also functions as a permissive. The turbine receives energy to turn from the boiler steam header. Hydraulic operated throttle valves control the amount of steam admitted to the turbine. The throttle valves are designed to fail-closed and stop all steam from entering the turbine on a loss of control oil. When in the open position, the interlock valve drains control

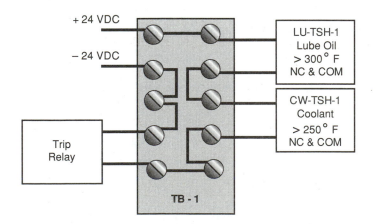

Figure 9.3-2 *Hardwired Trip String*

190

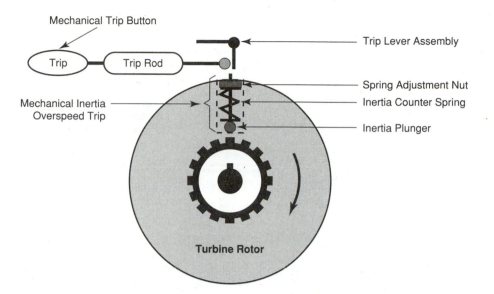

Figure 9.3-3 *Mechanical Trip Assembly*

oil back to the reservoir and prevents the control oil header from pressurizing. The interlock valve must close to build enough control oil header pressure to open the throttle valves. The interlock valve is similar to a fail-open, air-operated valve, except instead of using pressurized air to close, lubricating oil pressure is applied to the actuator. Therefore, the interlock valve prohibits turbine operation unless adequate lubricating oil pressure

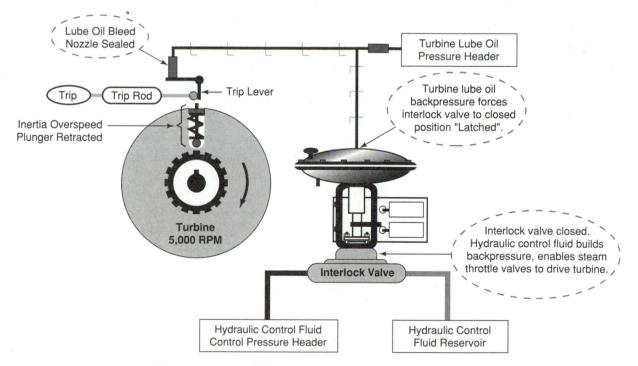

Figure 9.3-4 *Overspeed Reset/Turbine Latched*

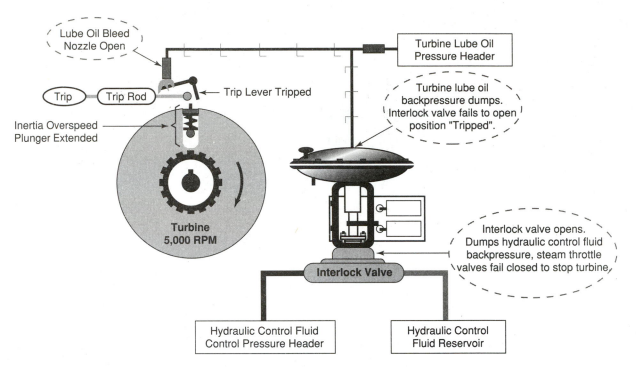

Figure 9.3-5 *Overspeed Tripped/Turbine Tripped*

is available. In Figure 9.3-5, a restrictive orifice supplies bearing lubricating oil to the intercept valve and the trip lever bleed nozzle. The orifice is sized to provide less oil than the bleed can vent. If the turbine over-speeds, the plunger extends, engages the trip lever, and vents bearing lubricating oil from the interlock valve actuator. As pressure at the actuator decreases the interlock valve opens and depressurizes the control oil header. Once control oil pressure drops, the throttle valves fail-close, the trip/throttle valves shut off steam to the turbine, and the turbine stops.

Figure 9.3-6 displays redundancy attained by installing a true dual diverse redundant electrical overspeed trip system. The system monitors turbine speed by electronically sensing the frequency of pulses produced by a toothed wheel attached to the turbine shaft. The speed controller compares actual speed to maximum allowable speed set-point. If either speed probe senses excess speed, the controller interrupts electrical power to both solenoid dump valves. One solenoid fails-open and dumps bearing oil from the interlock valve actuator. This forces the interlock valve to fail-open and dump control oil pressure. The other solenoid fails-open and dumps control oil directly to the reservoir. This type of system is extremely reliable, especially if true redundant power supplies are provided to the electrical overspeed controller. A magneto is sometimes attached to the machine's shaft as a permanent generator to provide electrical power to the control circuits as long as the shaft is spinning. Some turbines have shaft driven lubricating oil pumps that provide the primary bearing lubricating oil. In this case, the back-up lubricating oil pump is electrically driven and only used during start-up or if the primary shaft driven pump fails. The safeguards applied in this example are extensive; they are expensive to design, install, and maintain.

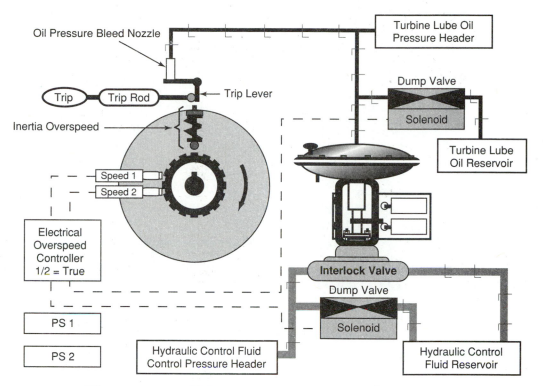

Figure 9.3-6 *Triple Redundant Overspeed Protection*

This is, however, a realistic representation considering the destructive forces involved in the obliteration of a large turbine due to overspeed.

Chapter Nine Summary

- The purpose of emergency shutdown systems is to:
- Eliminate unnecessary trips or transients due to single point failures.
- Reduce or mitigate undesirable effects of human performance errors.
- Provide a Safe and Controlled response to unexpected conditions.
- Protect equipment from damage.
- Protect the environment from contamination.
- Prevent personal injury.
- Installation of redundant instrumentation is used by ESD systems for securing the process.
- Control valves are configured for specific fail-safe positions and particular failure modes.
- Failure mode selections include fail-as-is, fail-open, or fail-closed.
- Fail-open valves require air to close on top of the diaphragm, fail-close valves require air to open on the bottom of the diaphragm, and fail-as-is valves accept air to the top and bottom of the diaphragm.

- Fail modes are designed to configure or restore the process to the safest or most desirable condition in the event of a loss or degradation of control system capabilities.
- Additional fail-safe support devices include solenoids, springs, accumulators, and shuttle valves.
- Redundant systems use several identical hardware channels to simultaneously monitor the same point.
- The extent of safeguards incorporated is determined by the volatility of the process and the negative impact resulting from a loss of control.
- One or more techniques are employed, based on engineering review of design-based impact assessments, industry events, and plant operating experience.
- True fail-safe systems return the process to a controlled condition without dependency on electrical, pneumatic, or hydraulic power.
- Trips automatically interrupt or reconfigure final control device(s) if the measured variable exceeds specifications.
- Trip actuations typically fail control valves by shutting off air supply via a solenoid and/or de-energizing electrically operated valves (MOVs).
- The purpose of interlocks and permissives is to allow equipment to start and operate only when monitored variables are within designed parameters.
- Typical process variables monitored by interlocks, trips, and permissives are flow, level, temperature, pressure, chemistry, contamination, and reactivity.
- Typical machine variables monitored by interlocks, trips, and permissives are coolant level, temperature, and pressure; lubricant level, temperature, and pressure; and machine vibration, temperature, and speed.
- Interlock designs are mechanical, electrical hardwired (relay logic), electrical software (processor controlled), pneumatic, or hydraulic.

Chapter Nine Review Questions:

1. State the purpose and intent of emergency shutdown systems.

2. List three fail-safe modes of a valve.

3. Identify a fail-open valve.

4. Identify a fail-close valve.

5. Identify a fail-as-is valve.

6. State the purpose of fail-safe valves.

7. List the fail-safe support devices.

8. Define the term single point failures.

9. Define the term mitigate.

10. State the fundamental requirement of a fail-safe system.

11. Describe several methods used by ESD systems for securing the process.

12. Define redundancy.

13. State the benefits of redundant controls.

14. State the drawbacks of redundant controls.

15. State three methods used to determine if redundant control is required.

16. State the purpose of interlocks and permissives.

17. State the purpose of trips.

18. List the process variables typically monitored by interlocks, trips, or permissives.

19. List the machine variables typically monitored as interlocks, trips, or permissives.

20. List the motive medias used to effect interlocks, trips, and permissives.

Chapter 10
Programmable Logic Controllers

OBJECTIVES

Upon completion of Chapter Ten the student will be able to:

- State the purpose and function of a programmable logic controller (PLC).
- List the fundamental hardware components of a PLC.
- State the function of the individual hardware components of a PLC.
- List the typical inputs and outputs monitored by a PLC.
- Convert binary numbers to octal and hexadecimal equivalents.
- Identify ladder logic symbols.
- List typical logic operands used by a PLC.
- Interpret ladder logic diagrams.
- Identify the hardware termination input/output point specified by the ladder logic.
- Determine if the anticipated field input is recognized by a PLC.
- Determine if a module contains a faulty fuse.
- Determine if the interface module is communicating.
- Determine if the power supply is OK.
- Locate and interpret self-health status monitoring LEDs.
- State the conditions required to generate a discordance alarm.

10.0 Introduction

Programmable logic controllers (PLCs) combine extremely high computation ability, diverse flexibility, and excellent reliability into a space-efficient control system. Their use is widespread throughout industry. PLCs control livestock feed and watering systems, lathes and milling machinery, water purification, and a host of assembly line operations. Batch processes in chemical, pharmaceutical, and food /beverage industries all rely on PLCs. Many skid-mounted process systems are delivered with preprogrammed PLCs already installed. Since PLCs have the ability to communicate with other processors, they are easily added to existing systems. They adapt readily to design alterations and are reconfigured as required to meet emerging needs.

10.1 Programmable Logic Controller Hardware

Programmable logic controllers are microprocessor-based sequential step software configurable controllers. Typical hardware, shown in Figure 10.1-1, consists of a mounting rack, power supply, comunication interface module, processor module (CPU), and a selection of user-defined plug in input/output modules. The mounting rack provides a stable-mounting platform for the modules. It also contains the back-plane bussing network that distributes power and data through the rack. The power supply provides back-plane power to the modules and signal power or wetting voltage for the output modules. The

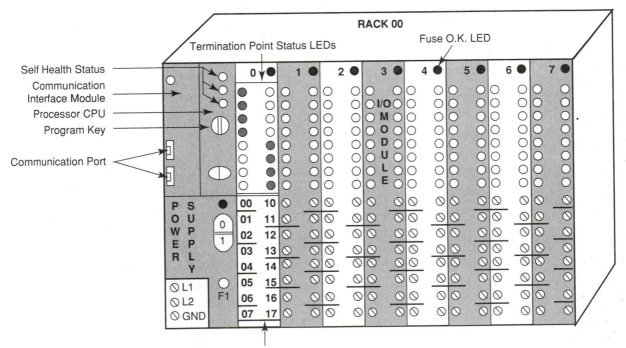

Figure 10.1-1 *Programmable Logic Controller Hardware*

communication interface module provides data transport links to other racks in the system, and RS232 serial or RJ11 phone connections to facilitate user programming with a laptop computer or handheld communicator. The processor module (CPU) provides memory locations for executive programs, application programs (ladder logic), data bits, and registers. The processor performs all logic functions and executes the application program. Input/output modules receive and transmit data to and from field devices. They provide electrical isolation to protect the CPU and transpose data into the appropriate retransmission format. The type of data transmitted determines input/output module selection. Specific modules are selected to input or output either binary or analog data. Typical input modules are installed to monitor AC volts binary input, DC volts binary input, DC volts analog input, DC current analog input, thermocouple analog input, and RTD analog input. Typical output modules are AC volt binary, DC volt binary, and DC volts analog. Multiple rack configurations are capable of sharing input data through the communication interface module. The PLC monitors external contacts, performs logical computation and comparison to the user installed program code, and outputs the desired data. The installed program code is accessible through the communication module and is revised or updated as required. The program software is presented in a user friendly, ladder diagram format.

10.2 Binary Numbering Systems

Programmable logic controllers only manipulate data in binary format. Since long strings of ones and zeros are difficult for people to work with, alternate numbering systems are used to represent groups of three and four bits of binary data as alphanumeric symbols. The alternate numbering systems used are binary (base two), octal (base eight), and hexadecimal (base sixteen). Figure 10.2-1 lists the equivalent values of these numbering systems through 15_{10}. To avoid confusion multiple based numbering systems are identified by subscript: decimal$_{10}$, binary$_2$, octal$_8$, and hexadecimal$_{16}$. Since multi-digit non-decimal numbers represent different values than the decimal value of the same symbols, different terms are used. The octal value 13_8 is referred to as "one three octal," not thirteen. The hexadecimal term 13_{16} is referred to as "one three-hex." Additionally, when raising numbers to powers or using exponents, any number raised to the power of zero is "one." Any number raised to the power of one is unchanged. Numbers raised to powers other than zero (0) or one (1) are multiplied by themselves the number of times indicated by the power to which they are raised. Some examples are $2^0 = 1$, $2^1 = 2$, $2^2 = 4$, and $10^3 = 1000$.

The universal numbering system in use today is the decimal system. The decimal system is base ten. This means there are only ten symbols from which to select. The symbols are 0, 1, 2, 3, 4, 5, 6, 7, 8, and 9. These ten symbols are arranged in sequence to represent any numeric value. This works well as the count progresses from 0 to 9. However, when the count reaches 9, all available symbols are used. To progress the count, place weighting is required. A 0 is used to indicate the one's column is full and the count is resumed in the ten's column. Therefore, any decimal number is actually a summation of the counts in each column, with each column representing an additional multiple of ten.

Decimal	Binary (Four Bit)	Octal	Hexadecimal
0	0000	0	0
1	0001	1	1
2	0010	2	2
3	0011	3	3
4	0100	4	4
5	0101	5	5
6	0110	6	6
7	0111	7	7
8	1000	10	8
9	1001	11	9
10	1010	12	A
11	1011	13	B
12	1100	14	C
13	1101	15	D
14	1110	16	E
15	1111	17	F

Figure 10.2-1 *Numerical Conversion*

Example: The number 1942 represents,

A count of "1" in the one thousand column $(1 \times 10^3) = 1 \times 1000 = 1000$

Plus a count of "9" in the one hundred column $+ (9 \times 10^2) = 9 \times 100 = 900$

Plus a count of "4" in the ten column $+ (4 \times 10^1) = 4 \times 10 = 40$

Plus a count of "2" in the one column $+ (2 \times 10^0) = 2 \times 1 = 2$

The summation of these counts is 1942_{10}.

Decimal, binary, octal, and hexadecimal numbering systems all adhere to the same place weighting principle.

The binary system is base two. This means there are only two symbols from which to select. The symbols are 0 and 1. These two symbols are arranged in sequence to represent any numeric value. This works well as the count progresses from 0 to 1. However, when the count reaches 1, all available symbols are used. To progress the count, place weighting is required. A 0 is used to indicate the one's column is full and the count is resumed in the two's column. Therefore, any binary number is actually a summation of the counts in each column, with each column representing an additional multiple of two.

Example: The number 1101_2 represents,

A count of "1" in the eight column $= (1 \times 2^3) = 1 \times 8 = 8$

Plus a count of "1" in the four column $+ (1 \times 2^2) = 1 \times 4 = 4$

Plus a count of "0" in the two column $+ (0 \times 2^1) = 0 \times 2 = 0$

Plus a count of "1" in the one column + $(1 \times 2^0) = 1 \times 1 = 1$
The summation of these counts = 13_{10}.

The octal system is base eight. This means there are eight discrete symbols from which to select. The symbols are 0, 1, 2, 3, 4, 5, 6, and 7. These eight symbols are arranged in sequence to represent any numeric value. This works well as the count progresses from 0 to 7. However, when the count reaches 7, all available symbols are used. To progress the count, place weighting is required. A 0 is used to indicate the one's column is full and the count is resumed in the eight's column. Therefore, any octal number is actually a summation of the counts in each column, with each column representing an additional multiple of eight.

Example: The number 243_8 represents,
A count of "2" in the sixty-four column = $(2 \times 8^2) = 2 \times 64 = 128$
Plus count of "4" in the eight column + $(4 \times 8^1) = 4 \times 8 = 32$
Plus a count of "3" in the one column + $(3 \times 8^0) = 3 \times 1 = 3$
The summation of these counts is = 99_{10}.

The hexadecimal system is base sixteen. This means there are sixteen discrete symbols from which to select. The symbols are 0, 1, 2, 3, 4, 5, 6, 7, 8, 9, A, B, C, D, E, and F. These sixteen symbols are arranged in sequence to represent any numeric value. This works well as the count progresses from 0 to F. However, when the count reaches F, all available symbols are used. To progress the count, place weighting is required. A 0 is used to indicate the one's column is full and the count is resumed in the sixteen's column. Therefore, any hexadecimal number is actually a summation of the counts in each column, with each column representing an additional multiple of sixteen.

Example: The number $13C_{16}$ represents,
A count of "1" in the two hundred fifty-six column = $(1 \times 16^2) = 1 \times 256 = 256$
Plus a count of "3" in the sixteen column + $(3 \times 16^1) = 3 \times 16 = 16$
Plus a count of "C" in the one column + $(C \times 16^0) = C \times 1 = C$
The summation of these counts is = 316_{10}.

Binary numbers are often represented as groups of three or four. Three-bit binary numbers convert readily with octal, and four-bit binary numbers convert readily with hexadecimal. Always begin the grouping with the least significant bit, 2^0. If the last grouping contains less than the three or four bits needed for conversion, zeros are placed in the empty high order columns to complete the group.

Example: Group this binary number into three-bit binary format. 10111011_2
Begin with the lowest order bit and separate the bits into groups of three. $10 - 111 - 011_2$
The high-order group of three contains only two of the desired three bits.
Place a zero in the vacant highest order column $010 - 111 - 011_2$.

The zero has no effect to the value of the term, but preserves the three bit-grouping formats.

Example: Group this binary number into four-bit binary format. 101011100_2
Begin with the lowest order bit and separate into groups of four. $1 - 0101 - 1100_2$
There are less than the required four bits in the higher-order term.
Place zeros in the vacant high-order columns. $0001\ 0101\ 1100_2$.

Again, the zeros have no effect on the term, but merely preserve the four-bit grouping formats.

Three full place weights of binary total to a sum of seven. Starting the count at zero provides eight counts. Therefore, a three-bit binary number can represent any single octal symbol. Each additional digit in the octal number requires an additional set of three binary bits.

Example: $7_8 = 111_2$ and $17_8 = 001\ 111_2$
Conversely, to convert any binary number into octal, group the binary digits in groups of three and convert each three-bit group into the octal equivalent.

Example: 101001111_2 separated into groups of three is $101\ 001\ 111_2$
The highest order group converts to 5_8
The next group converts to 1_8
And the low-order group converts to 7_8.
Therefore, the octal number is 517_8.

Four place weights of binary total to a sum of fifteen. Beginning the count with zero provides sixteen counts. Therefore, a four-bit binary number can represent any single hexadecimal symbol. Each additional digit in the hexadecimal number requires an additional set of four binary bits.

Example: $F_{16} = 1111$ and $3C_{16} = 0011\ 1100$.

Conversely, to convert any binary number into hexadecimal, group the binary digits into groups of four and convert each four-bit group into the hexadecimal equivalent.

Example: 101001111_2 separated into groups of four is $0001\ 0100\ 1111_2$.
The highest order group converts to 1_{16}
The next group converts to 4_{16}
And the low order group converts to F_{16}
Therefore, the hexadecimal number is $14F_{16}$.

10.3 Ladder Diagrams

Ladder diagrams are line code illustrations of programmable logic controller, process control software. Ladder diagrams document the manipulation and transmission of internal and

Mnemonic	Instruction	Mnemonic	Instruction
ADD	add	AVE	average
CMP	compare	CTD	count down
CTU	count up	DIV	divide
GEQ	greater than or equal to	LEQ	less than or equal to
MCR	master control reset	MUL	multiply
NEQ	not equal to	OTE	energize
OTL	latch	OTU	unlatch
PID	proportional/integral/derivitive	RES	reset
TOF	timer off delay	TON	timer on delay

Figure 10.3-1 *Typical Mnemonics*

external data. Rungs contain memory addresses, program code, or Boolean commands. The address for the device is a wiring block termination point, memory location, or a contact on a different rung. Unlike line drawings that only represent external contact data of actual hardware components, ladder diagrams represent internal (software) and external (hardware) contact data. Memory addresses, program codes, and Boolean commands provide internal data. Field contacts from a pressure, temperature, level, flow, or a valve position switch provide external data. Contacts from pushbuttons, hand switches, relays, and timers also provide external data. Rung outputs drive alarms, stroke control valves, start/stop motors and pumps, transmit binary data, and provide logic states back to other circuits.

Ladder diagrams are illustrated similar to line drawings, except the main lines (rails) on the ladder diagram are drawn vertically and the individual circuits (rungs) are placed horizontally. The input/bus and output/bus are considered as rails of the ladder. Each subordinate circuit is an individual rung on the ladder. Each rung contains a condition instruction (input) and a command instruction (output). Each successive rung on the ladder is sequentially acknowledged. As the condition instruction is realized (made true), the command instruction on that rung is executed. The functional description and address identify symbols on each rung. The type of instruction is identified as I (input), O (output), T (times), or C (counter). Figure 10.3-1 lists the actual instruction as identified by a mnemonic symbol assigned by the manufacturer.

10.4 Input/Output Addressing

Instruction addresses are presented in binary, octal, or hexadecimal format. The first set of numbers identifies the physical rack. The second set identifies the physical module, and the third set identifies the actual (external) or virtual (internal) terminal. The actual address for the device is a wiring block termination point. Figure 10.4-1 depicts the virtual address memory mapped to the specific data bit in the word address representing the particular hardware module. The octal number I0/007/12 locates the termination point in rack (cabinet) 000, input module (card) 7, and terminal (screw) 12, shown in Figure 10.4-2. This

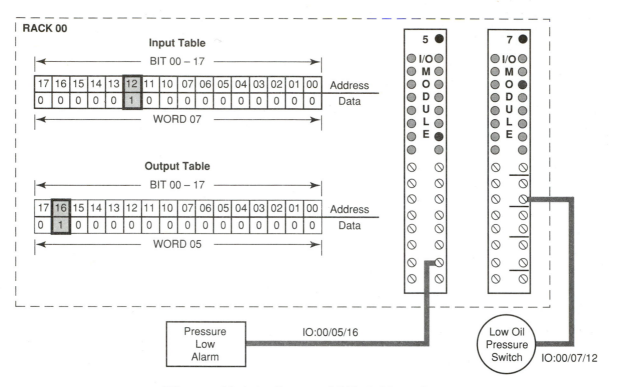

Figure 10.4-1 *Internal I/O Addressing*

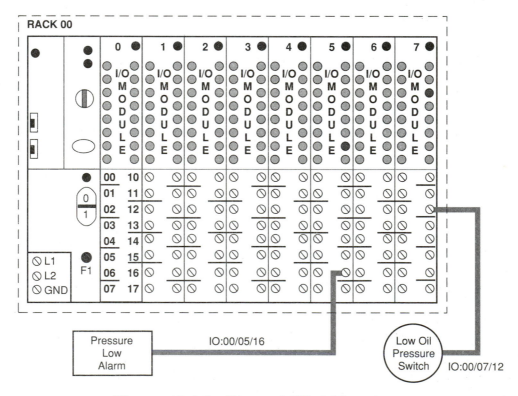

Figure 10.4-2 *External I/O Addressing*

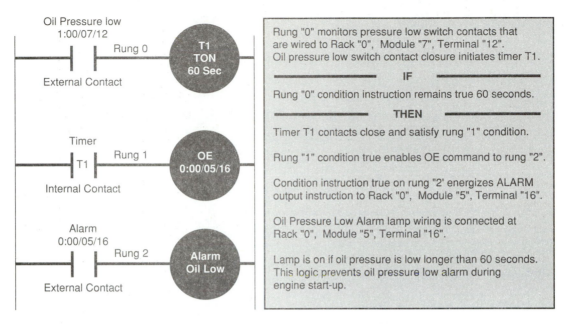

Figure 10.4-3 *Ladder Logic Diagram*

data point is mirrored in the memory map at word I0/00/7 bit 12. The ladder diagram in Figure 10.4-3 depicts logic that prevents the low oil pressure alarm from coming on until the oil pressure is low for longer than sixty seconds. This type of logic is also used to prevent nuisance alarms from noise, spikes, and transient conditions. See Figure 10.4-4.

Figure 10.4-4 *Programmable Logic Control Block*

10.5 Troubleshooting

Termination point status lights are provided on each binary I/O module. If the condition at any particular termination point is true, the light illuminates. If the condition monitored at the termination point violates the program sequence, a discordance alarm is generated, and the status light blinks until the discordance is eliminated. A multimeter is used to monitor the input/output terminals and verify the LED status indication is correct. Exercise caution and verify the multimeter is configured to monitor the value expected. The module type indicates the expected value. Typical values are 12 – 120 VAC, 12 – 120 VDC, or 4 – 20 mADC. Internal fuses at each I/O termination point protect the processor from excess current. The I/O modules have a single blown fuse indicator that illuminates if any fuse in the module fails to conduct. Fuse replacement may require partial removal of the affected module. Never remove or install an I/O module until the rack is de-energized. The interface and processor modules also display current internal status update data. They each provide several LEDs to indicate system health and activity. A local communication port is often provided on the processor. A handheld communicator or laptop computer containing the appropriate software is used to read input status, read output status, or toggle an output point. A section of memory is usually dedicated to store error flags or error bits. The communicator is used to view this information. The processor module is provided with a read/write/run key switch to prevent accidental or unauthorized changes to the software. Most PLC related failures are actually failed field devices providing erroneous information or failing to actuate on demand.

■ Chapter Ten Summary

- Programmable logic controllers are microprocessor-based sequential step software configurable controllers.
- The hardware consists of a mounting rack, power supply, communication interface module, processor module, and various application-specific input/output modules.
- The mounting rack provides a stable platform for the modules, and contains the back-plane bussing network that distributes power and data through the rack.
- The power supply provides back-plane power to the modules and signal power or contact wetting voltage for the output modules.
- The communication interface module provides data transport links to other racks in the system, and RS232 serial or RJ11 phone connections to facilitate user programming with a laptop computer or handheld communicator.
- The processor module (CPU) provides memory locations for executive programs, application programs (ladder logic), data bits, and registers. It also performs all logic functions and executes the application program.
- Input/output modules receive and transmit data to and from field devices, provide electrical isolation to protect the CPU, and transpose data into the appropriate retransmission format.

- The type of data transmitted determines input/output module selection. Typical input modules are AC volts binary input, DC volts binary input, DC volts analog input, DC milliamps analog input, thermocouple analog input, and RTD analog input. Typical output modules are AC volt binary, DC volt binary, and DC volts analog.
- Ladder diagrams are line illustrations of process control application software.
- Ladder diagrams are represented as two parallel vertical lines connected by horizontal rungs. Rungs depict inputs from and outputs to field devices, including the specific address of each data point.
- The functional description and address identify symbols on each rung. The type of instruction is identified as I (input), O (output), T (times), or C (counter). The actual instruction is identified by a mnemonic symbol assigned by the manufacturer.
- Each rung contains a condition instruction (input) and a command instruction (output). As each successive rung on the ladder is sequentially acknowledged, the condition instruction is realized (made true), and the command instruction (mnemonic) on that rung is executed.
- Contacts from pushbuttons, hand switches, relays, timers, pressure, temperature, level, flow, or valve position switches provide external data.
- Rung outputs drive alarms, stroke control valves, start/stop motors and pumps, transmit binary data, and provide logic states back to other circuits.
- The address of the instruction is presented in binary, octal, or hexadecimal format.
- The first set of numbers identifies the physical rack; the second set identifies the physical module; and the third set identifies the actual (external) terminal. The actual address for the device is a wiring block termination point. The virtual address is memory mapped to the specific data bit in the word address representing the particular rack and module.
- Termination point status lights are provided on each binary I/O module. If the condition at any particular termination point is true, the light illuminates.
- If the condition monitored at the termination point violates the program sequence, a discordance alarm is generated and the status light blinks until the discordance is eliminated.
- Internal fuses at each I/O termination point protect the processor from excess current. The I/O modules have a single blown fuse indicator that illuminates if any fuse in the module fails to conduct.
- The interface and processor modules also display internal status update data. They each provide several LEDs to indicate system health and activity. The processor module is provided with a read/write/run key switch to prevent accidental or unauthorized changes to the software.

Chapter Ten Review Questions:

1. State the purpose and function of a programmable logic controller (PLC).

2. List the five fundamental hardware components of a PLC.

3. Describe the function of each individual hardware component of a PLC.

4. List the typical inputs/outputs monitored by a PLC.

5. Convert 001 110 binary into octal and hexadecimal equivalents.

6. Draw ladder logic symbols for a normal-open contact and a normal-closed contact.

7. List typical logic operands used by a PLC.

8. Draw a simple switch alarm ladder logic diagram.

9. Locate the termination point represented by input table word I0/005/bit12.

10. State the method to determine if the anticipated field input is recognized by the PLC.

11. State the method to determine if a module contains a faulty fuse.

12. State the method to determine if the interface module is communicating.

13. State the method to determine if the power supply is O.K.

14. State the location of the self-health status monitoring LEDs.

15. State the conditions required to generate a discordance alarm.

Chapter 11

Distributive Control Systems

- *State three methods of human interface with the bus.*
- *List several methods used to display process data by the application software.*
- *State the purpose of the polling process as it relates to distributive control.*
- *State the definition of system topology as it relates to distributive control.*
- *State the definition of a network as it relates to distributive control.*
- *Develop simple diagrams depicting star, multidrop, and ring topology.*
- *Explain the relationship between traffic volume and bus capacity.*
- *Explain the relationship between line length, signal strength, and noise.*
- *Differentiate between introduced noise and induced noise.*
- *List the most common problems to the bus.*

11.0 Introduction

Previous chapters disclose the operation and function of simple local control loops. Typical loops contain a transmitter, controller, final control element, and the necessary indicators and alarms required to monitor the process. If the process deviates sufficiently an alarm is generated and the operation technician can locate the controller and adjust the set-point. Control systems that require the physical presence of an operator to oversee and manipulate the equipment are called local control systems. Local control is a reasonable application for intermittently operated equipment, extremely stable processes, and certain independent subsystems. There are several advantages of local control. Since all control is independent and localized at the device, monitoring functions are continuous, control reactions to process upset are immediate, and the control function capability is independent of a support station. However, complex processes require interdependent control actions, and a centralized monitoring and control station is required. Centralized control systems consist of a rack-mounted card frame controller, a programmable logic controller, or a host central processing unit that performs automatic control calculations and algorithms, and a main control station with indicators and alarms, and the support controls needed to remotely reposition final elements, change set-points, and start/stop pumps and motors.

Central control stations are often called control rooms, and the processor area is called the rack room. Control rooms and rack rooms are climate controlled, access-restricted areas. The entire unit or plant system is graphically represented on control boards or panels that line the walls. Critical switches, indicators, and alarms are located on the panel where the associated device is displayed. A human machine interface (HMI), a keyboard and monitor display, perform this same function in digital systems. Process systems are grouped on the panel in a manner that supports logical interpretation of the indications and alarms.

Distributive control systems (DCS) combine the advantages of centralized monitoring systems and localized control. A distributive control system is an interactive network of instrumentation, communication, and control devices connected by a common communication link. Process algorithms are calculated and executed locally at the field device, and process variable, machine condition, and other critical parameters are monitored at a local HMI or at a central station. Distributive control systems with intelligent instrumentation also monitor self-health and perform limited diagnostics.

11.1 Architecture

Panel-mounted controllers serve multiple functions in pneumatic applications of centralized control. They perform control algorithm calculation and manipulation (PID), provide the HMI functions of indication and recording process data, and also provide the pushbuttons and switches required to tune parameters or initiate manual control. Individual tubing runs transmit the process variable from the local transmitter back to the control board, the controller indicates and records the value on a spool of graph paper, calculates a correction, and transmits the correction signal back to the final control element via a tubing. Cascade and other interdependent control strategies are mapped by routing the tubing from one controller to the other. Instrument air is routed throughout the plant and each instrument receives supply pressure through tubing. Tubing connections are often labeled at the device. Supply is labeled (S), process signal in is labeled (E), and final device control signal out is labeled (V). See Figure 11.1-1.

Centralized control for analog electronic equipment is a similar architecture. The significant difference is that the operational supply and the signal data are transmitted over the same medium, in this case twisted pair wire. All field cabling interfaces with the control rack at the termination panel. The termination panel is a series of terminal strips or connectors mounted to a panel. Good designs use a consistent terminal identification pattern

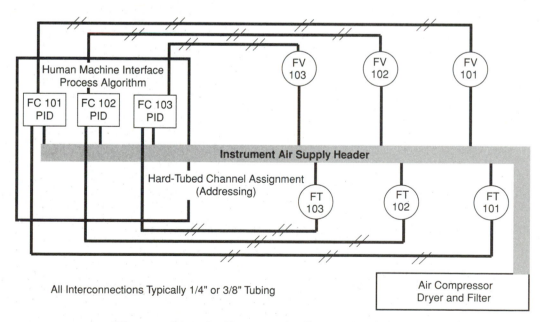

Figure 11.1-1 *Pneumatic Centralized Control*

throughout the entire rack. A card frame houses the individual controller cards and provides loop power to all field components. A variety of function cards are available, and the control strategy is mapped by selectively wire wrapping individual connections between the termination panel and the card frame backplane. Connector cables route the signals from the backplane to the selected card in the frame. Each function card has edge-mounted conductors that fit into the connector when the card is inserted into the rack. Function cards may have as few as eight or as many as forty individual conductor pins. Most rack-mount cards offer multiple function capability and are often field configured for the desired operation. Card-mounted jumper or dipswitch position determines auto/manual, direct/reverse, PID, cascade, and other operational criteria. When replacing cards it is important to verify that the revision of the replacement card is suitable for the intended functions, and that all jumpers and dipswitches are in the desired position. Specialized cards in the rack support control room indication, alarm, and set-point controls. Hybrid electronic analog control, such as HART, is identical in architecture to traditional electronic analog. It does, however, add the ability to remotely monitor and service field equipment. See Figure 11.1-2.

Distributive control systems are actually a combination of three different technologies: intelligent (microprocessor-based) instruments, communication networks (data links), and application software. Many applications incorporate H1 fieldbus at the plant floor level and

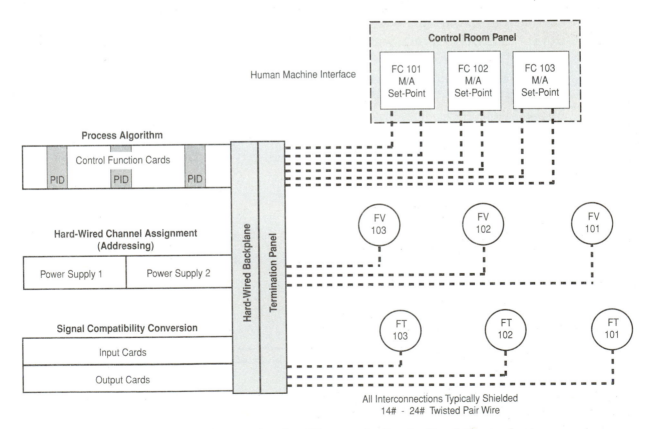

Figure 11.1-2 *Analog Electronic Centralized Control*

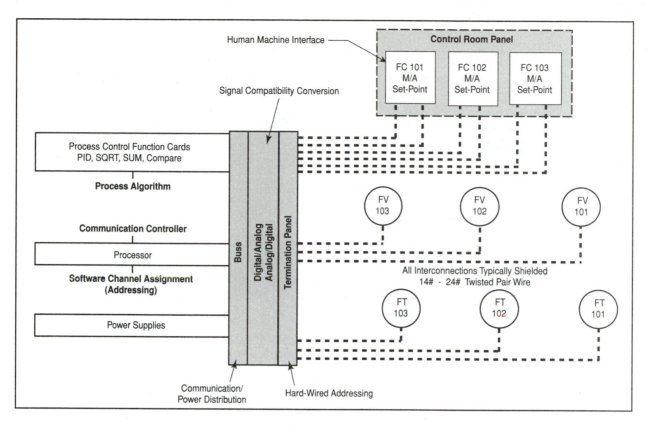

Digital Electronic Distributed Control

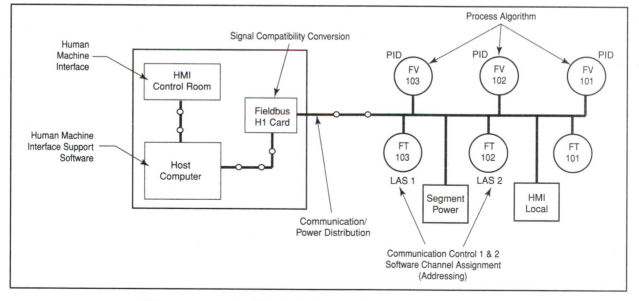

Figure 11.1-3 *Digital Control System Architecture*

use high-speed ethernet (HSE) for plant-wide distribution. The actual process control is determined locally as the devices on each fieldbus segment execute their individually assigned functions and algorithms. One device on each segment is a designated link active

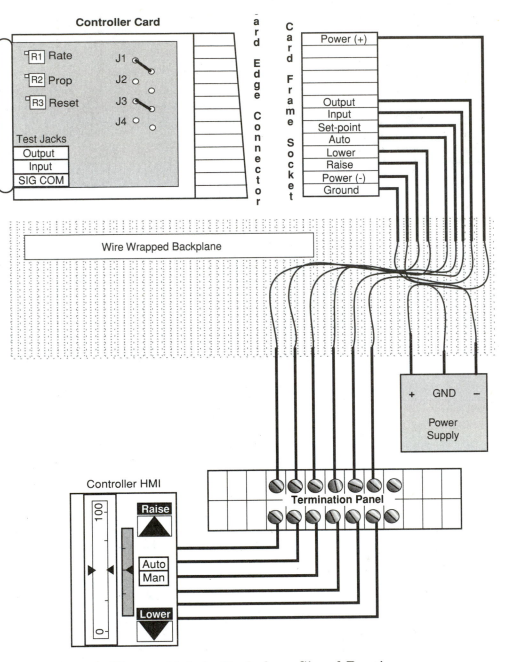

Figure 11.1-4 *Backplane Signal Routing*

scheduler (LAS). The LAS controls communication between devices on the segment and performs operability and self-health functions that maintain segment integrity. Fieldbus segments continue to operate independently and maintain process control functions without direct support of a host computer or an HMI. Interface devices are installed to connect devices to bus segments and facilitate cross protocol communication between busses. High-speed ethernet is often used to link fieldbus segments, and to link the plant floor level of control and data acquisition to the control room and various HMI terminals. This allows the host computer to configure control room graphic displays, data logging, and annunciation. Many applications support a third echelon of data interchange by linking the HSE layer to a local area network (LAN). Once available to the LAN, plant process conditions and device parameters are accessed through conventional communication protocols via modem. See Figure 11.1-3. Figure 11.1-4 depicts hardwired backplane signal routing.

11.2 Hardware

Distributive control system hardware consists of a master control device such as a PLC or host CPU, an array of intelligent monitoring and control instruments (transmitters, controllers, and valves), various peripheral devices, such as a remote operator interface CPU, display, and keypad, Figure 11.2-1, and the cabling and communication interface components that create the data link.

The master control unit, or host, contains the supervisory control and data acquisition display and control (host application) software and regulates network communication. It

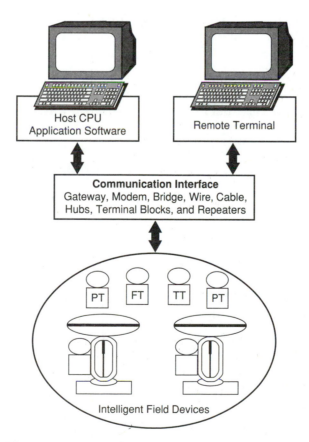

Figure 11.2-1 *Distributive Control System*

offers the highest level of user interaction. Program uploads and downloads, configuration changes, and actual active process commands are generally only authorized at the host CPU. Intelligent instrumentation devices such as transmitters, controllers, and control valves are mounted in the field to measure and control the process variables. An HMI that consists of a processor, display screen, and keypad allows the operator to view current process and equipment condition, request historical data, initiate trending of particular points, and verify field device status.

A bus is a communication interface that consists of cabling, terminations, power supplies, and necessary interface modules. The main bus, data hi-way, is often called the trunk line. Communication links from the field back to the control room are often called home runs. Busses connect network devices together and permit the transmission and reception of data. Each addressed device on the bus is a node. Conductors are usually twisted pair shielded wire, coaxial cable, or multiconductor cable. H1 low power fieldbus is compatible with conventional 4 – 20 twisted pair conductors, but high speed ethernet requires category five RG-45 cables. Special T- connectors and node modules physically connect devices to the bus. Individual remote power supplies are often installed to provide regulated operating voltage to the instruments on each section of the bus. Power conditioner modules are used to provide correct impedance and filtering. Power supply voltage is specified as 9 – 32 volts and typically 21 – 32 volts. Terminators are installed at the end of each drop to stabilize the signal, reduce line noise, and prevent ringing. Terminators are resistor/inductor circuits tuned to match the cable impedance. Amplifiers and repeaters are installed at critical points along the bus to maintain signal integrity and increase the useful length of the runs. Amplifiers boost the incoming signal and send it further down the bus. Amplifiers have negligible effect on transmission speed, but any distortion or noise present in the original signal is also amplified and retransmitted. Repeaters sample the original incoming signal and regenerate a duplicate signal for retransmission. Signal quality is maintained but transmission speed is affected. Other interface devices are communication gateways, I/O modules, bridges, and scanners.

11.3 Software

Protocol is a set of rules that govern the composition and transmission of binary data. Binary communication is transmitted as a string of ones and zeros. Protocol functions to segregate the ones and zeros into specific ordered groups or frames of serial data that are readily interpreted by devices on the bus. Serial data strings represent a complete and intelligent statement or command. Synchronized clock pulses establish a common point of reference for the binary data. Each serial transmission contains a designated start and stop code. Addresses, commands, and data are each assigned a specific ordered sequence within the message, and the type of error checking is identified. Foundation fieldbus uses manchester encoding. See Figures 11.3-1 and 11.3-2.

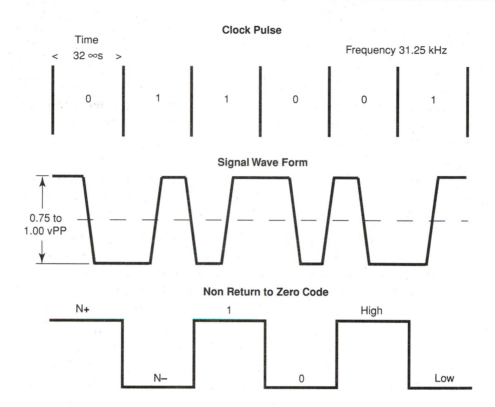

Clock Pulse

Time
< 32 ∞s >
Frequency 31.25 kHz

0   1   1   0   0   1

Signal Wave Form

0.75 to
1.00 vPP

Non Return to Zero Code

N+   1   High

N–   0   Low

Each Fieldbus Device Typically Draws 10 mA to 20 mA at Steady State.

Device Toggles Current Draw by 0.25 to 0.33 mA.

0.25 to O.33 mA Current Times 3 K Ohm Power Supply Impedance = 0.75 to 1.00 vPP.

1 (One) ---------- Hi-Lo Transition (Mid-bit)

0 (Zero) ---------- Lo-Hi Transition (Mid-bit)

N+ (Non-data plus) ---------- Hi (No Transition)

N– (Non-data minus) ---------- Lo (No Transition)

Figure 11.3-1 *Manchester Encoded Fieldbus Signal*

Intelligent instruments contain an independent programmable processor with memory, and have on-board self-diagnostic and communication capability. On-board processors contain packets of information or blocks. Each device has a single resource block that contains specific information unique to the hardware of the individual instrument. Resource block information includes manufacturer's identification number, device type and revision number, memory size and space available, available features, and the device state. Transducer blocks read and write sensor input/output and isolate the function blocks.

A single device may contain several transducer blocks. Transducer blocks are used to calibrate the device, check sensor health, and perform on-line sensor diagnostics. Function blocks contain the configuration and control parameters that establish the function of the device and execute the desired control strategy. The key to fieldbus distributed control is the ability of each device to implement control strategy and process algorithms independently in the field. Common function blocks and their primary function are listed in Figure 11.3-3. Each function block contains a set of configurable parameters, Figure 11.3-3. Software driver files called device descriptions are available for foundation fieldbus devices. They include all common parameters and functions of each device type. Device descriptions are

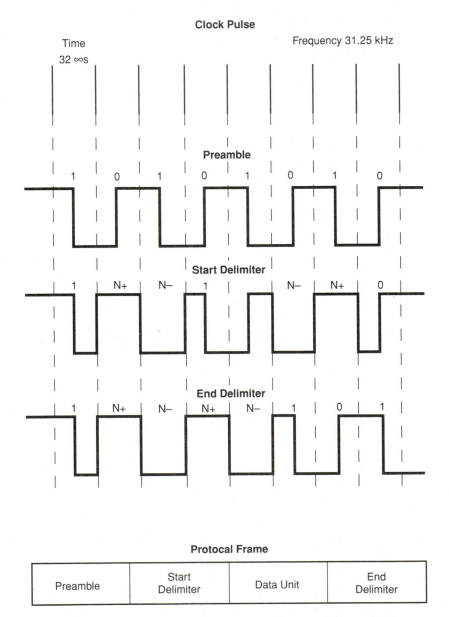

Figure 11.3-2 *Physical Layer Protocol Data Unit*

AI	Analog In	Accepts digitized external analog signal (reads analog input). Performs scaling, filtering, high alarm, and low alarm.
AO	Analog Out	Provides digitized analog signal (sends analog output). Includes scaling, range, rate limits, and fault state options.
DI	Discreet In	Accepts 8-bit input value from external hardware (reads discrete input). Provides filtering, inversion, and discrete alarms.
DO	Discrete Out	Passes 8-bit output value to external hardware(sends discrete output). Provides filtering, mode shedding, and fault state.
PID	Proportional, Integral, Derivative Controller	Provides filtering, setpoint limits,feedforward signal, output limits, error alarms, mode shedding, and anti-windup (PID control).
PD	Proportional + Derivative	Provides filtering, setpoint limits,feedforward signal, output limits, error alarms, mode shedding, and anti-windup (P+D control).
ML	Manual Loader	Accepts digitized analog signal from an AI block or computer. Provides high and low output limits and an output to other blocks (manual control).
BG	Bias/Gain	Provides calculation with output limits to other blocks (scales).
CS	Control Selector	Provides high, low, average select from several inputs (override control).
RA	Ratio Station	Accepts signal from two AI blocks and computes setpoint for controller block to control ratio.

Figure 11.3-3 *Basic Fieldbus Function Blocks*

used to construct and configure the control strategy. Application software uses tag data polling addresses to identify and verify each particular instrument. Process variable updates as well as control and configure command communications are relegated to a particular address.

At this point, it may be beneficial to review fieldbus in relation to conventional control strategies. A typical level control loop might include a level transmitter (LT-100), a level indicator (LI-100), a level high switch (LSH-100), a level low switch (LSL-100), a level controller (LC-100), and a level control valve (LCV-100) with valve position switches (ZSC-100) and (ZSO-100). Each of these devices is manufactured to perform limited specific functions. They all support some type of scale and operability tuning but the basic function of each device is fixed by the inherent design of the device. Since all components are individually hardwired, data transfer routing is permanently fixed: The transmitter can only transfer analog process value to the indicator and the controller. The level switches can only transfer discrete alert data to the annunciation panel. The control valve accepts an analog position demand from the controller, and the valve position limit switches transmit discrete valve position data to the control panel valve position indication lights. Since each hardware device is limited in function, a simple P&ID is usually a sufficient representation of the loop and the applied control strategy. Logic diagrams depict the functionality of the device in relation to the loop. The logic diagram does not represent the actual components of the loop, but defines the control strategy as a series of logical operations. These logical operations are annotated as blocks, with each block performing a specific function, and nand, or, nor, not, and latch.

A typical fieldbus level control loop includes significantly less hardware. Each device is capable of a high degree of interoperability. Various function blocks within the device are assigned to perform a variety of operations. Since all devices within each segment are wired to the same twisted pair cable, functions can be assigned and located within any device on the loop. All functions of the previous hardware are accomplished with only a level transmitter (LT-100) and a level control valve (LCV-100). The transmitter analog input function block places the digitized process variable on the bus segment. Discrete input blocks within the transmitter place level high/level low alerts on the bus segment. The control valve PID block performs the controller function and the control valve discrete input block provides valve position data to the bus. Field-bussed loops are represented by function block configuration diagrams, see Figure 11.3-4. The control strategy is mapped with function block diagrams that are similar in concept to logic diagrams. Each function block is also identified with the actual field device physical tag number within which the function block resides. Although no recognized standards are accepted at this time, in current practice fieldbus devices are displayed on P&ID drawings using conventional device symbols with the characters FF annotating them as fieldbus devices. The active function blocks are

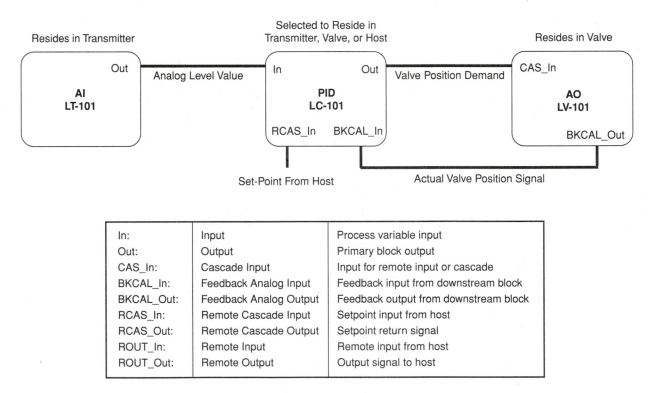

Figure 11.3-4 *Simple Level Loop Function Map*

drawn as attached squares affixed to the device in which they reside; computer interlink symbols portray the bus cabling.

Handheld remote communicators can interface with the distributive control system at any point on the bus. Communicators use instrument tag data to identify and communicate with all intelligent instruments on the bus. The full array of applications, however, is not usually accessible with the handheld.

Application software performs several primary functions. It administers and regulates cross-bus communication, data transfer, and storage. It monitors and controls process variables. It implements self-health and diagnostic analysis, and supports human machine interface. Application software contains a "virtual" process system that mimics the actual hardware. The control software utilizes instrument addresses to send and receive data from each instrument on the data link. As actual field devices provide new information, the software updates the virtual representations of the device, Figure 11.3-5. Supervisory control and data acquisition software collects, analyzes, and updates data from the monitored points. It generates graphic displays, trends, plots, and graphs of process variable values, and automatically processes actuations, notifications, and alarms. The program also monitors overall system health. It initiates a polling process that performs routine instrument operability checks and verifies communication integrity of the bus. Software also supports operator interface capability. Additional application software functions available through the host or remote terminal include data log recovery/display, cross-calibration

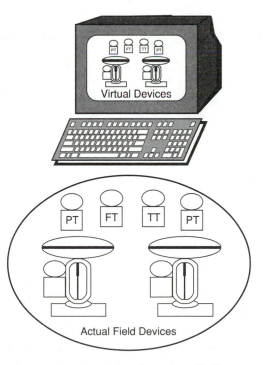

Figure 11.3-5 *Application Software*

comparisons, first-out event analysis, flow chart programming, troubleshooting, corrective action statements, diagnostic analysis, and predictive maintenance schedules.

11.4 Topology

Distributive control systems are configured with various topologies. System topology refers to the particular technique employed to physically connect individual devices throughout the network. A network is a group of field and control devices connected through a common medium. Each addressed device connected to the network is a node. Common topologies are star, multidrop, and ring. Star topology connects several point-to-point devices to a central communicator, Figure 11.4-1. Multidrop topology connects all nodes in parallel to a single cable, Figure 11.4-2. Ring topology electrically connects the nodes in a series loop. Ring topology uses each device on the ring as a transceiver and repeater, Figure 11.4-3. Communication passes through all devices and only the specific devices addressed respond to the command. Combination topology is quite common, and many times results as additional devices and controls are incorporated into the original design, Figure 11.4-4.

Many distributive control malfunctions are self-diagnosed and self-corrected. If a particular instrument or node experiences a failure, the master controller sets an alert flag and removes that device or group of devices from service. If parallel channels are available the host transfers local control to them. Certain programs substitute an averaged signal in place of missing process data until the failure is corrected. Most distributive control network

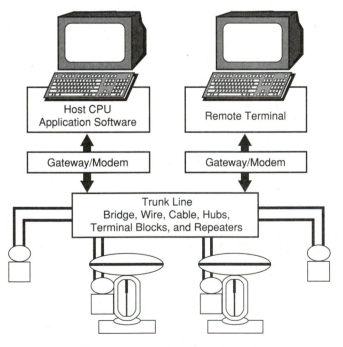

Figure 11.4-1 *Star Topology*

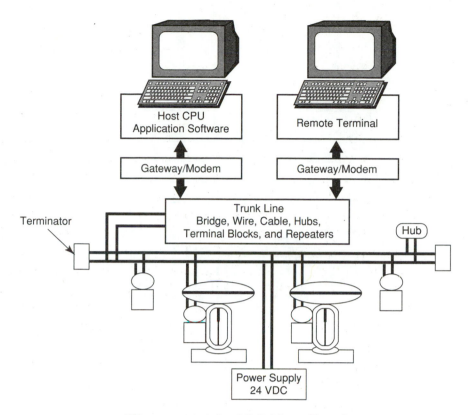

Figure 11.4-2 *Multidrop Topology*

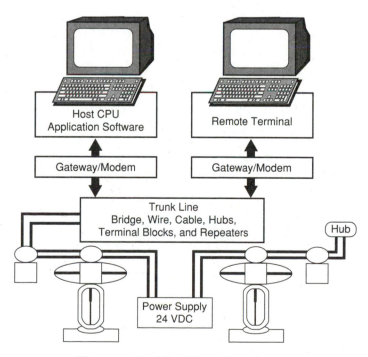

Figure 11.4-3 *Ring Topology*

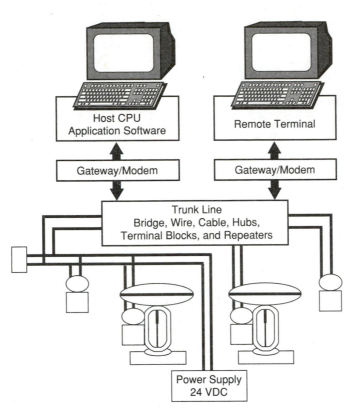

Figure 11.4-4 *Combined Topology*

problems appear shortly after the system is placed in service. Data collisions occur when several different devices attempt to transmit simultaneously. With several different devices talking at once, it is difficult for the devices that are listening to make much sense of it all. A single device can create similar fault conditions on a fieldbus segment if it continually transmits out of sequence. In these instances, it is difficult for the scheduler to deactivate the device because the communication link is saturated with jabber. High traffic evolutions such as system start-up can stress a system that is already loaded to capacity and induce problems that are not evident under normal conditions. Excessive line length increases noise and attenuates the data signal. Grounding problems, power supply failures, and noise are the most common problems to the bus.

11.5 Operability Assessment

Control system operability depends on the integrity of each individual device, as well as the pressure/power supply and the signal distribution medium. Pneumatic systems require clean, dry, filtered air of sufficient volume and pressure to operate. Loop requirements are calculated prior to construction. The total air consumption of each device and the combined length of the tubing runs are used to determine the required pipe/tubing diameter, and the compressor/receiver size. Initial loop checks prior to, or during, system start-up verify the operability of the instrument air system. Each tube run is bubble-tested to verify the absence of leaks. Rotameters are installed at each device on the loop and adjusted to simulate maximum rated flow and pressure is measured to validate minimum required pressure at maximum rated flow. Good design incorporates sufficient margin to support minor additions but major modifications require reevaluation of air supply requirements. Desiccants beak down over time and care must be taken to ensure fine particulates do not pass through the after filter and enter the piping. Proper maintenance includes main air supply desiccant monitoring and replacement, moisture trap/dryer blow down, and filter replacement. Each device should have an individual filter regulator attached. Filter regulators should be routinely blown down and adjusted.

There are several techniques to determine the integrity of bus cabling. They do, however, require specific skill and knowledge. Incorrect use of measuring and test equipment can result in personal injury and additional damage to bus devices. Only trained personnel should perform the following system checks. Isolate all measuring and test equipment connected to the bus from plant ground. Use battery-powered devices, or ensure the proper isolating transformer is used for AC powered equipment. Always determine and maintain correct polarity. A multimeter connected to measure ohms or amperes initiates a direct short, current path, through the meter. To verify correct cabling, de-energize and isolate the bus. To isolate a short circuit condition, measure the resistance of each conductor to ground with a digital multimeter. A high resistance indicates an open condition and a low resistance indicates a short. Power and signal conductors should read practically infinite resistance (> 20 M ohms) to ground. The shield and drain conductor should read practically zero resistance to ground when terminated and infinity when the single point ground is lifted. All conductors should read high resistance (> 50 K ohms), from one to the other, and practically infinite resistance (> 20 M ohms) to the shield. To verify continuity of the conductors, sequentially and temporarily tie each conductor to ground at one end of the run and measure conductor resistance to ground from the other end of the run. A good conductor should read extremely low resistance (<10 ohms). High ohms resistance indicates a damaged or cross-wired conductor. Disconnect the temporary ground before proceeding to the next conductor. Continue until all conductors are verified, Figures 11.5-1 and 11.5-2. Several instrument companies offer fieldbus specific test equipment. These devices inject a specified signal onto the segment at one end of the bus, and monitor the signal for strength and clarity at the furthermost point on the bus. Other devices clip to the bus segment and determine signal strength, power supply voltage, and noise by monitoring existing transmissions. Acceptable signal strength is 0.80 – 1.2 volts P-P. The inadvertent addition of extra terminators tends to attenuate (reduce) the signal strength; the inadvertent omission of a terminator tends to amplify (increase) signal strength. Power supply voltage should measure between 9 and 32 VDC at any point on the bus.

Power supply problems include low voltage, low current, and noise. To investigate the power supply, disconnect the load and measure the output voltage with a multimeter. If the voltage is incorrect, replace the power supply. If the voltage is within specified tolerances, monitor current and voltage as the load is reintroduced. Low voltage and/or high current

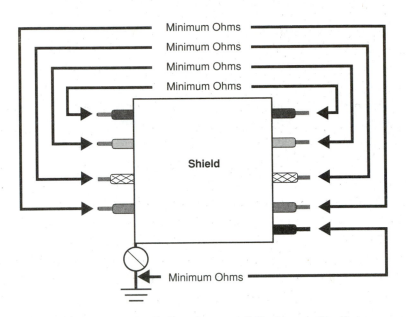

Figure 11.5-1 *Ideal Conductors/Minimum Resistance*

indicate an excessive load, but not necessarily a ground fault. Compare the current requirements of the combined devices on the bus to the power supply specifications. If possible, measure voltage at several points along the bus and attempt to locate any excessive voltage drops. Noise can reduce signal integrity, especially where line length is excessive. Devices on the bus, particularly power supplies can introduce noise, and high-energy electromagnetic fields can induce noise onto the bus. To determine noise on a DC power supply, measure for millivolts AC at the power supply output, and compare the measurement to manufacturer's specifications. Induced noise often occurs as an intermittent spike and is difficult to document even with recorders and oscilloscopes. Visually inspect the cable runs for proximity to high-energy devices or activities. High

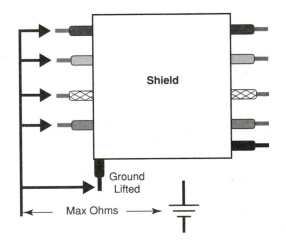

Figure 11.5-2 *Floated Conductors Isolated*

voltage motor controls, welding, and radiography can induce noise on the bus. Re-route the cable or increase cable shielding in these areas. Cycle suspected equipment and attempt to correlate data transmission faults to the operation of specific equipment.

The calibration technique for centralized pneumatic loops is basically a harmonized collection of individual component calibrations. The particular advantages include the ability to inject a measured process variable at the primary sensor, transmitter, and document additive error as it progresses through the loop. Additional advantages include the ability to test the interconnecting tubing for channel assignment, leaks, and plugging. Air consumption issues are also more readily disclosed when all components are monitored simultaneously under dynamic conditions. Certain scaling issues that might pass unnoticed through independent device calibration are easily identified during loop calibrations. Figure 11.5-3 demonstrates typical signal injection and monitoring points.

Analog card frame control systems are also calibrated similar to a combination of device calibrations. As with pneumatic loops, advantages include power supply monitoring under load, interconnecting cabling verification, and scaling confirmation. Most electronic analog card frame control systems perform signal conditioning and control computation within the

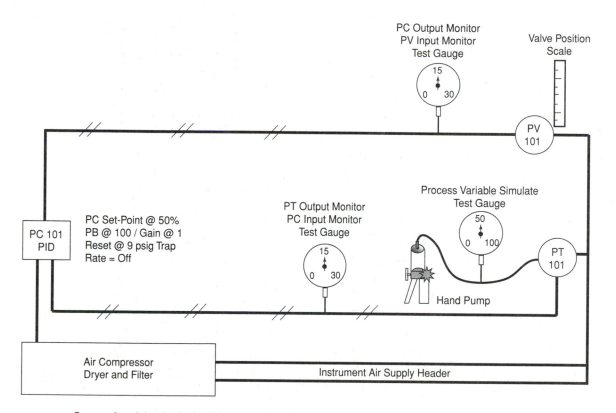

Once as found data is obtained, begin adjustments at transmitter, and complete calibration at valve. Highest degree of certainty is obtained by simultaneous monitoring and calibration with entire loop. Driven from process variable simulation device.

Figure 11.5-3 *Pneumatic Control Loop Calibration*

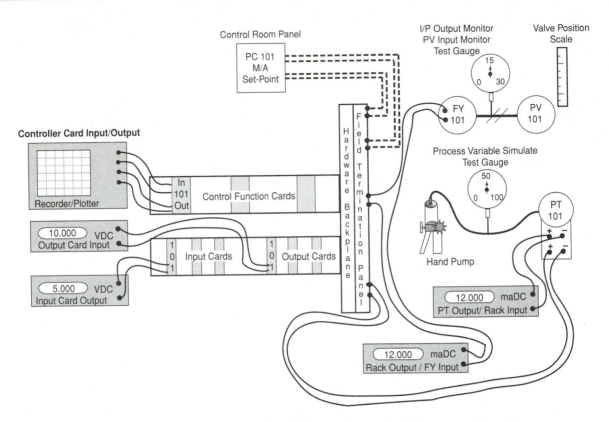

Figure 11.5-4 *Analog Card Frame Loop Callibration (Static)*

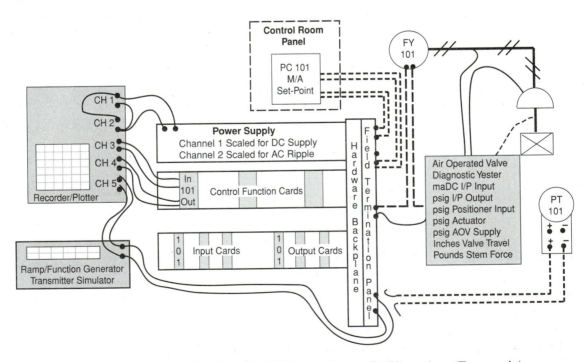

Figure 11.5-5 *Analog Card Frame Loop Callibration (Dynamic)*

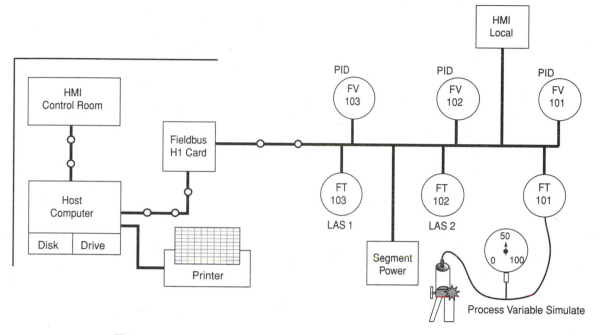

Figure 11.5-6 *Digital Electronic Distributed Control*

control cards. Input/output and control cards usually provide test jacks or test point pins to facilitate point-to-point monitoring of rack components. Figure 11.5-4 demonstrates a typical static calibration set-up. The process variable is incrementally varied in both an increasing and decreasing direction, and the signal is allowed to stabilize prior to recording data. Figure 11.5-5 depicts a typical dynamic calibration set-up. The transmitter is removed from the loop and replaced with a ramp generator. The generator is configured to inject a ramp symbolic of process dynamics and all relevant points are plotted in real time. Careful review of the graph/plot is required to reveal actual loop response. Control dynamics are often calculated in response to a defined step change. Figure 11.5-6 depicts a typical digital system calibration. The values are available at multiple monitoring points and injected data is automatically recorded. Observe polarity, ground, and isolation requirements of the loop when installing test equipment.

■ Chapter Eleven Summary

- A distributive control system is an interactive network of instrumentation, communication, and control devices connected by a common communication link.
- Distributive control systems are actually a combination of three different technologies, intelligent (microprocessor-based) instruments, communication networks (data links), and application software.
- Distributive control system hardware consists of a master control device such as a PLC or host CPU, an array of monitoring and control instruments (transmitters, controllers,

and valves), various peripheral devices such as a remote operator interface CPU, display, and keypad, and the cabling and communication interface components that create the data link.

- The master control unit contains the display and control (host application) software and regulates network communication.
- Transmitters, controllers, and control valves are mounted in the field to measure and control the process variables.
- An operator interface (terminal) that consists of a processor, display screen, and keypad allows the operator to view current process and equipment condition, request historical data, initiate trending of particular points, and verify field device status.
- A bus is any communication interface consisting of cabling, terminations, power supplies, and necessary interface modules.
- Busses connect network devices together and permit the transmission and reception of data.
- Each addressed device on the bus is a node.
- Conductors are usually twisted pair shielded wire, coaxial cable, or multiconductor cable.
- Special T- connectors and node modules physically connect devices to the bus.
- Individual remote power supplies are often installed to provide regulated operating voltage to the instruments on each section of the bus.
- Terminators are installed at the end of each drop to stabilize the signal and reduce line noise.
- Amplifiers and repeaters are installed at critical points along the bus to maintain signal integrity and increase the useful length of the runs.
- Interface devices are installed to connect devices to bus segments and facilitate cross-protocol communication between busses.
- Protocol is a set of rules that govern the composition and transmission of binary data. Protocol segregates the ones and zeros into specific ordered groups or frames of serial data that is readily interpreted by devices on the bus.
- Each serial transmission contains a designated start and stop code. Addresses, commands, and data are each assigned a specific ordered sequence within the message.
- On-board processors contain a packet of information specific to each individual in strument called tag data.
- Tag data includes the polling address, manufacturer's specification, calibration data, and tuning algorithms unique to the particular device.
- Application software uses tag data polling addresses to identify and verify each particular instrument.
- Handheld remote communicators interface with the distributive control system from any point on the bus.
- Communicators use instrument tag data to identify and communicate with all intelligent instruments on the bus.
- Application software performs several primary functions. It administers and regulates bus communication, data transfer, and storage. It monitors and controls process

variables. It implements self-health and diagnostic analysis, and supports human machine interface.

- System topology refers to the technique employed to physically or electrically connect individual devices throughout the network. Common topologies are star, multidrop, and ring.

- A network is a group of field and control devices connected through a common medium.

- Most distributive control network problems appear shortly after the system is placed in service.

- High traffic evolutions such as system start-up can stress a system that is already loaded to capacity and induce problems that are not evident under normal conditions.

- Grounding problems, power supply failures, and noise are the most common problems to the bus.

Chapter Eleven Review Questions:

1. State the definition of a distributive control system.

2. List the three different technologies that comprise a distributive control system.

3. List the basic hardware components used in a distributive control system.

4. State the purpose of the master control unit.

5. State the purpose of field-mounted instrumentation devices.

6. State the purpose of an operator interface (terminal).

7. State the purpose of the bus as it relates to distributive control.

8. State the definition of a node as it relates to distributive control.

9. List several types of conductors used in distributive control busses.

10. Identify components used to physically connect devices to the bus.

11. State the purpose of individual remote power supplies on the bus.

12. State the purpose of terminators as they relate to distributive control.

13. State the purpose of amplifiers and repeaters as they relate to distributive control.

14. State the purpose of interface devices as they relate to distributive control.

15. State the function and purpose of protocol.

16. List the information included in instrument tag data as it relates to distributive control.

17. State three methods of human interface with the bus.

18. List several methods used to display process data by the application software.

19. State the definition of a network as it relates to distributive control.

20. Explain the relationship between traffic volume and bus capacity.

21. Explain the relationship between line length, signal strength, and noise.

22. Differentiate between introduced noise and induced noise.

Chapter 12

Practical Application of Knowledge

OBJECTIVES

Upon completion of Chapter Twelve the student will be able to:

- *List steps required before performing any equipment manipulations.*
- *State the common method of mounting tank level instruments.*
- *State the adverse effects of manipulating tank level instruments mounted on a common standpipe.*
- *State the proper valve sequences to remove and restore level instruments to service.*
- *State the reason to use caution when venting or draining a level instrument.*
- *State the adverse effects of manipulating instruments that share common sensing lines.*
- *State two sensing line conditions that create significant measurement error.*
- *State a condition to avoid when manipulating differential pressure isolation valves.*
- *State the normal valve sequences to remove and restore a transmitter to service.*
- *State the proper sequence to execute an auto/manual transfer.*
- *State the proper sequence to execute a manual/auto transfer.*
- *State the proper sequence to remove a valve from service.*
- *State the proper sequence to restore a valve to service in auto control.*
- *State the proper sequence to restore a valve to service in manual control.*
- *State the three common types of control response.*

- *Describe quarter wave dampening (ultimate period).*
- *Apply the Ziegler/Nichols calculation.*
- *Describe minimum disturbance tuning.*
- *Describe minimum area tuning.*
- *Describe the relationship between reset output and valve position.*
- *Describe the relationship between derivative output and valve position.*
- *Describe the relationship between proportional output and valve position.*
- *Explain the dynamic control response of proportional (P).*
- *Explain the dynamic control response of reset (I).*
- *Explain the dynamic control response of rate (D).*

12.0 Introduction

The following information is represented as a fundamental approach or guide only. Only trained and qualified technicians with extensive hands-on training for each particular loop should attempt field tuning or manipulating active process systems and control loops. Before any equipment manipulations, establish communications between the board and field operators to review all affected control schemes, develop guidelines, and establish monitoring of the process variable.

12.1 Removing and Restoring Instruments to Service

Tank level instruments are often connected to a common standpipe as in Figure 12.1-1. Any change or disturbance to the standpipe inlet/outlet valves will affect all instruments attached to the common standpipe. Manipulating level switch and level control isolation valves can unexpectedly change the standpipe level and result in unanticipated alarms and actuations. Level instruments typically have a high-side inlet valve and a vent valve installed on the upper inlet, and low-side isolation valve and drain valve connected to the lower inlet. Fully close both isolation valves before opening the vent or drain valves. It is critical to remember that even with both upper and lower isolation valves closed, full process pressure is still trapped in the pot. Use extreme caution when opening the vent and/or drain.

Transmitters are often removed and restored to service. Although it is poor practice, instruments may share common sensing lines. If several components are connected, as in Figure 12.1-2, special precautions are required to prevent upsets. Manipulating transmitter isolation valves can induce spikes on the sensing lines, which could result in an unanticipated alarm or actuation. Many instruments are installed with filled sensing lines. Flow transmitters are extremely sensitive to sense line conditions. They require a sense line that is completely dry and devoid of all moisture, or completely solid and devoid of all air. A loss of sensing line reference fluid, or entry of air into sealed systems, creates significant measurement error. Long runs of sensing line are installed at a slope, with vent and drain valves installed to facilitate air removal. To prevent damage to the sensor the transmitter

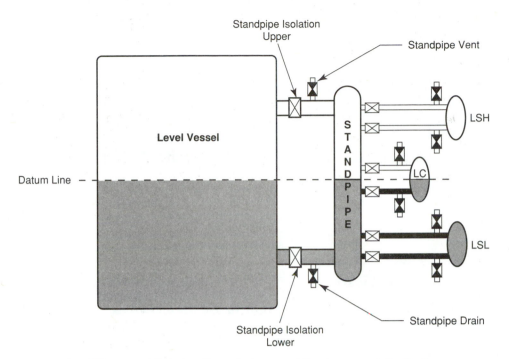

Figure 12.1-1 *Standpipe and Instrument Isolations*

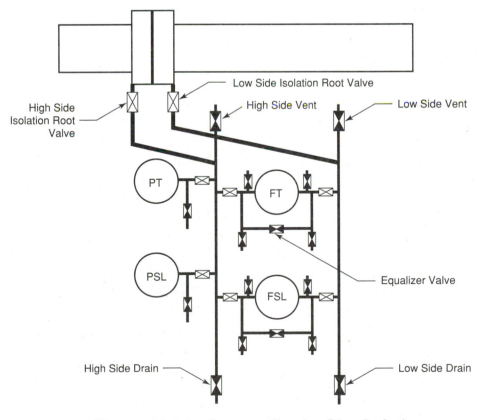

Figure 12.1-2 *Common Sensing Line Isolation*

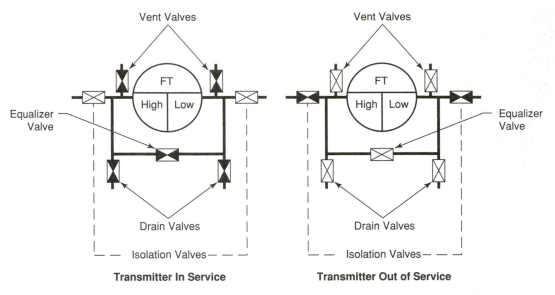

Figure 12.1-3 *Transmitter Isolation Valves*

is never exposed to excessive differential pressure. The traditional method to isolate a transmitter is to first close the high-side process inlet valve, refer to Figure 12.1-3, then open the transmitter equalization valve. This ensures that the low-side inlet pressure is applied equally to both sides of the transmitter. Once the transmitter is equalized, close the low side inlet valve. Transmitters are restored to service in a similar manner. First, open the transmitter equalization valve. Then, open the low-side inlet valve. Allow sufficient time for the transmitter to equalize and close the transmitter equalization valve. Open the transmitter high-side inlet valve and restore the transmitter to service. Vent and fill the transmitter and sensing lines.

> **CAUTION** **Vent and fill activities potentially expose process media to the environment. Catch all fluid in a suitable container and dispose of properly.**

Ensure proper precautions are employed when venting and filling high-energy systems.

The transmitter is provided with two venting screws, one for each side. Determine and implement the required precautions for the particular process fluid. Slowly loosen each screw to allow a minute amount of fluid to vent. Exercise caution to ensure the transmitter is not exposed to excessive differential pressure. Aerated fluid appears opaque and exits the vent in erratic spurts. Continue the venting until all air is removed from the sensing lines. If it is not practical or desirable to use the process for venting the sensing lines, isolate and disconnect the sense lines at the orifice root valves. Connect a pressurized fluid source to the transmitter drain valves and inject fluid until all air is forced out at the root valve connection. If unable to isolate and disconnect the sensing lines, monitor the pressure value of the pressurized fluid as it is injected. When adding additional fluid to the sense line ceases to increase the monitored backpressure, the line is full.

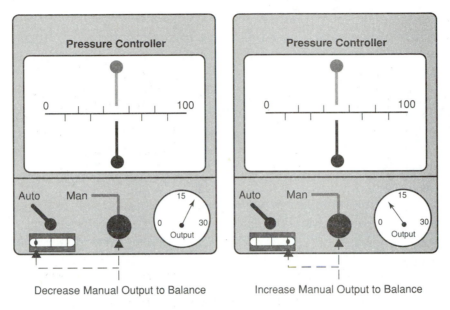

Figure 12.2-1 *Auto/Manual Transfer*

12.2 Auto/Manual Transfer

Auto/manual transfer is often required to support maintenance activities and other evolutions. It is critical to execute a bumpless transfer, the transfer is transparent, and that the final control element remains stable. While in auto control, adjust the manual output to match the auto output. Once the outputs are equal, transfer to manual control and ramp the final element to the desired position. Adjustments are identified in Figure 12.2-1.

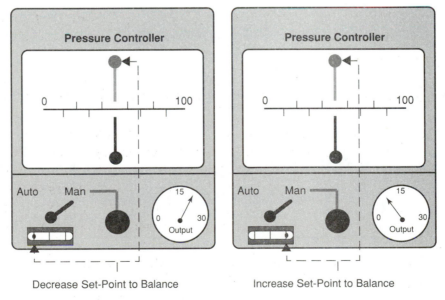

Figure 12.2-2 *Manual/Auto Transfer*

Manual/auto transfer is accomplished much the same way as auto/manual transfer. Manipulate the controller set-point adjustment to match the auto output signal to the manual demand signal. Once the outputs are equal, place the system in auto control and adjust the controller set-point as required to ramp the final control element to the desired position. Adjustments are identified in Figure 12.2-2. Always ensure that the automatic demand signal and the manual demand signal are equal before attempting the transfer.

12.3 Removing and Restoring Valves to Service

The preferred method for removing a valve from service is to maintain auto control and slowly throttle the by-pass valve into service. Observe the primary final control element move toward full closed as the by-pass valve assumes the flow. Once the primary valve reaches full closed, manipulate the controller set-point to overdrive the primary valve (hard seat). Set the manual control output to match the auto control output and transfer to manual control. Secure the primary final control element. This sequence is depicted in Figures 12.3-1, 12.3-2, and 12.3-3.

There are also different methods for restoring a valve to service. The preferred method is with the controller in automatic control. First, manipulate the controller set-point to match the auto output signal to the manual demand signal. Allow sufficient time, as auto response is delayed if the controller is in a reset wind-up condition. Once the demand signals match, place the controller in automatic. Slowly throttle the by-pass valve closed, and observe the primary final control element modulate open and into control as the by-pass valve assumes less flow. Adjust the controller set-point to ramp the primary valve to the desired control position.

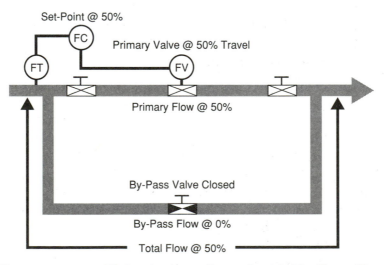

Figure 12.3-1 *Valve in Auto Control with By-Pass Closed*

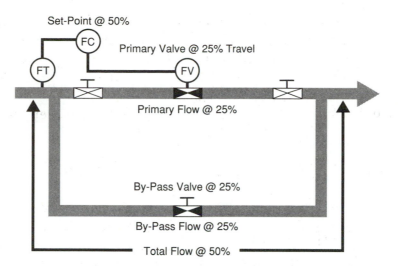

Figure 12.3-2 *Valve in Auto with By-Pass Throttled*

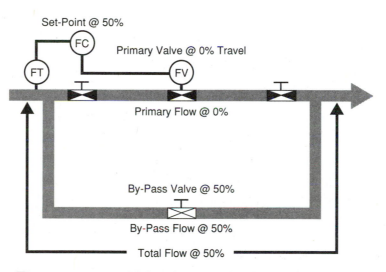

Figure 12.3-3 *Valve Secured By-Pass Controlling*

To restore a valve to service with the controller in manual control, ensure the controller manual demand signal matches the primary valve position. Slowly adjust the manual output to ramp the primary valve toward the desired position. Throttle the by-pass closed and verify that the controller auto signal responds to process conditions. Allow sufficient time, as auto response is delayed if the controller is in a reset wind-up condition. Set the manual control output to match the auto control output and place the controller in auto control. Adjust the controller set-point to drive the primary valve to the desired position.

12.4 Field Tuning Fundamentals

Optimum tuning of a control loop depends on the particular requirements and demands of the process monitored. Three common types of control response tuning are quarter wave (ultimate period), minimum disturbance, and minimum area. Quarter wave dampening (ultimate period) is the most common desired response. Quarter wave response is characterized by a decaying sinusoidal recovery response signal where each successive cycle attains only one-fourth-peak amplitude of the previous cycle. Quarter wave is highly effective at set-point recovery, as most disturbances are corrected in four cycles. There are, however, some applications where aggressive disturbance recovery is unacceptable. Minimum disturbance response is sometimes preferred. Minimum disturbance tuning produces high system stability. The process variable remains extremely stable, but recovery time is extended and some set-point offset often remains. Minimum area tuning is similar to quarter wave response. The desired effect of minimum area tuning is to develop a recovery response that permits the minimum deviant sinusoidal area. The minimum sinusoidal area represents an integration of process variable deviation from set-point over time. System response is rapid, but minor oscillations are sometimes present.

Many techniques are employed to restore control loop stability. The first step is to monitor the system to determine baseline conditions. Observe and record minimum/maximum deviation and cycle time and record the as-found controller settings. Compare the phase relationship between the measured variable and the valve position. Reset induced disturbances typically lag "out of phase" and the controller output continues to drive the valve past the optimum position. Once the base conditions are determined, judiciously and incrementally slow down the amount of controller reset. Allow several cycles after each adjustment. Compare the measured variable value to the set-point and, if required, correct the offset with the controller set-point control. If the system continues to oscillate with reset adjusted to the slowest value, other adjustments are required.

Observe the phase relationship between the final control element and the process variable. Derivative (rate) induced instability typically leads out of phase, and initially overdrives the valve beyond the optimum position. The magnitude of the step change in relation to the rate of change in the measured variable is derivative induced. Judiciously and incrementally reduce the controller rate adjustment. Allow several cycles after each adjustment. Gain/proportional band induced instability is generally in phase with the measured variable.

Increasing proportional band/decreasing gain increases system stability and decreases system responsiveness. Decreasing proportional band/increasing gain increases system response and reduces system stability. Judiciously and incrementally increase proportion band/decrease gain. Allow several cycles after each adjustment and, if required, correct offset with the controller set-point control. Once the process is stabilized, introduce an upset and monitor the effects. Increase rate to reduce dynamic deviation (not offset) and decrease recovery time. Speed up reset to eliminate any undesired offset (do not exceed cycle time).

In 1942, Ziegler and Nichols developed a technique that approximates quarter wave tuning. The basis of this technique determines the inherent sensitivity for each individual control loop and then applies proven formulas. The first step is to determine the ultimate gain and ultimate period for the control loop. First, remove all reset and derivative action from the controller. This is accomplished by setting integral to infinity and setting derivative to zero. With the controller in automatic, introduce a slight upset. Moving the set-point is usually sufficient. Observe the resultant waveform of the controller output. Adjust the gain as required to maintain a continuous steady-state oscillation at the controller output. If the wave decays, gain is too low. If the wave continues to increase in magnitude, the gain is too high. Once the gain is adjusted to create a sustained oscillation, record the gain value as ultimate sensitivity Su. Time the waveform for one period, peak to peak, and record this value as the ultimate period, Pu. The Ziegler/Nichols values are now applied to the loop values.

For a proportional-only controller $K_c = 0.5\ S_u$.

For proportional-plus-reset $K_c = 0.45\ S_u$ and $T_i = P_u\ /\ 1.2$.

For proportional-plus-derivative $K_c = 0.6\ S_u$, and $T_d = P_u\ /\ 8$

For proportional-plus-reset-plus-derivative $K_c = 0.6\ S_u$, $T_i = 0.5\ P_u$, and $T_d = P_u\ /\ 8$.

■ Chapter Twelve Summary

- Only trained and qualified technicians should perform tuning and manipulation of active process systems and control loops.
- Before performing any equipment manipulations, establish communications between the board and field operators to review all affected control schemes, develop guidelines, and establish monitoring of the process variable.
- Fully close both isolation valves before opening the vent or drain valves.
- Even with both upper and lower isolation valves closed, full process pressure is still trapped in the pot.
- Use extreme caution when opening the vent and/or drain.
- Manipulating transmitter isolation valves can induce spikes on the sensing lines that could result in an unanticipated alarm or actuation.
- A loss of sensing line reference fluid, or entry of air into sealed systems, creates significant measurement error.
- To isolate a transmitter, close the high side, open the equalization valve, and then close the low side valve.
- To restore a transmitter, open the equalization valve and then open the low side valve. Allow sufficient time for the transmitter to equalize and close the equalization valve. Open the transmitter's high-side valve and restore the transmitter to service.

- In auto control, adjust the manual output to match the auto output. Once the outputs are equal, transfer to manual control and ramp the final element to the desired position.

- To perform a manual/auto transfer, manipulate the controller set-point adjustment to match the auto output signal to the manual demand signal. Once the outputs are equal, place the system in auto control and adjust the controller set-point as required to ramp the final control element to the desired position.

- To remove a valve from service in automatic, maintain auto controls and slowly throttle the by-pass valve into service. Observe the primary final control element modulate toward full closed as the by-pass valve assumes the flow. Set the manual control output to match the auto control output and transfer to manual control.

- Three common types of control response tuning are quarter wave (ultimate period), minimum disturbance, and minimum area.

- Quarter wave response is characterized by a decaying sinusoidal recovery response signal where each successive cycle attains only one-fourth-peak amplitude of the previous cycle. Quarter wave is highly effective at set-point recovery, as most disturbances are corrected in four cycles.

- Minimum disturbance tuning produces high system stability. The process variable remains extremely stable, but recovery time is extended and some set-point offset often remains.

- The desired effect of minimum area tuning is to develop a recovery response that permits the minimum deviant sinusoidal area. System response is rapid but minor oscillations are sometimes present.

- Reset induced disturbances typically lag "out of phase" and the controller output continues to drive the valve past the optimum position.

- Derivative (rate) induced instability typically leads "out of phase" and initially overdrives the valve beyond the optimum position.

- Gain/proportional band induced instability is generally "in phase" with the measured variable.

- Increase rate to reduce dynamic deviation (not offset) and decrease recovery time.

- Speed up reset to eliminate any undesired offset (do not exceed cycle time).

Chapter Twelve Review Questions:

1. List the steps required before performing any equipment manipulations.

2. State the adverse effects of manipulating tank level instruments mounted on a common standpipe.

3. State the reason to use caution when venting or draining a level pot.

4. State the adverse effects of manipulating instruments that share common sensing lines.

5. State two sensing line conditions that create significant measurement error.

6. State the normal valve sequences to remove and restore a transmitter to service.

7. State the proper sequence to execute an auto/manual transfer.

8. State the proper sequence to execute a manual/auto transfer.

9. State the proper sequence to remove a valve from service.

10. State the proper sequence to restore a valve to service in auto control.

11. State the proper sequence to restore a valve to service in manual control.

12. Describe quarter wave dampening (ultimate period).

13. Describe minimum disturbance tuning.

14. Describe minimum area tuning.

15. Describe the phase relationship between reset output and valve position.

16. Describe the relationship between derivative output and valve position.

17. Describe the relationship between proportional output and valve position.

18. Explain the dynamic control response of reset.

19. Explain the dynamic control response of rate.

20. Explain the dynamic control response of proportional.

Chapter 13
Troubleshooting

OBJECTIVES

Upon completion of Chapter Thirteen the student will be able to:

- Define troubleshooting.
- State the expectation of troubleshooting.
- List the adverse effects of improper troubleshooting.
- State the difference between random failures and systematic failures.
- State the difference between permanent faults and random faults.
- List several possible sources of random faults.
- State several common techniques to isolate random faults.
- Describe the symptom recognition phase of troubleshooting.
- State the expectation of the symptom recognition phase of troubleshooting.
- List several resources for symptom recognition data.
- Describe the symptom analysis phase of troubleshooting.
- State the expectation of the symptom analysis phase of troubleshooting.
- Define a possible fault.
- Define a probable fault.
- Describe the system manipulation phase of troubleshooting.
- State the expectation of the system manipulation phase of troubleshooting.
- List several resources for system manipulation data.
- Describe the fault validation phase of troubleshooting.
- State the expectation of the fault validation phase of troubleshooting.
- List several resources for fault validation data.
- Describe the half split-system of fault validation.
- Describe the process to troubleshoot power supply and grounding faults.

13.0 Introduction

Troubleshooting is the ordered process of investigating, identifying, and eliminating a fault or combination of faults responsible for undesirable system conditions. Troubleshooting restores a system to desirable operating conditions and eliminates probable replicate failures. Proper troubleshooting techniques are critical to the safe and economical operation of any system. Troubleshooting is not the random replacement of parts until the symptoms of the undesirable condition are cloaked. Eliminating or masking symptoms without correcting the root cause creates a potentially dangerous and expensive situation. Intermittent failure of a pressure control device allows system pressure to momentarily exceed safe limits. If the recognized symptom is a high-pressure alarm, and the pressure switch that drives the alarm is found damaged, replacement of the pressure switch is obviously required. With the new switch in place, the alarm clears and the system appears to operate properly. An opportunity to identify and correct the real problem is not realized and the hazardous condition continues to exist. If maintenance history reveals that a particular device fails repeatedly, an underlying undesirable condition most likely exists.

Failures are either random or systematic. Random failures are spontaneous and generally hardware related. Systematic failures are caused by underlying faults of design or application. All faults are determined as systematic until proven random. Instrument devices are mass-produced. It is unlikely that a single device from the assembly line contains a degraded component. If one device exits the factory with a manufacturing fault there is a strong probability that many more devices left the factory in identical condition. Replacing a faulty power supply with an identical unit manufactured the same day may result in a short-term fix, but introduces a covert latent fault into the system. Faults are either permanent (hard) faults or random (intermittent) faults. A permanent fault creates adverse conditions that persist until the fault is remedied. Random faults create temporary adverse conditions and are difficult to locate. Random faults are often linked to other outside influences such as radio frequency transmission, cross talk, vibration, temperature, and moisture. Recorders and data loggers are quite helpful in catching random faults.

Problem solving requires the logical analysis of collected data. Successful problem solving depends on the proper analysis of complete and correct data. Compare and contrast similar systems with common faults, multiple monitoring points for common data, and system conditions before and after an event. Analyze cause and effect relationships to determine potential faults for a specific symptom and to relate or synchronize multiple symptoms to a single fault. Cause effect analysis is an important tool in problem solving. It is important to avoid post-hoc reasoning, such as assuming that since one event preceded another, the two events are directly related. Complete data includes current plant documents and maintenance history, equipment manufacturer's specification and service updates, user group, trade organization and governmental publications, and industry events. A systematic approach to problem solving has clearly defined expectations for each step of the evolution.

13.1 Symptom Recognition

The first phase of troubleshooting consists of symptom recognition and analysis. The primary expectation of symptom recognition is to determine the precise validity and identity of the undesirable condition. Most process control systems can be divided into actuation devices and indication/annunciation devices. The initial verification and validation process determines whether the undesirable condition is an actuation fault or an annunciation fault. An actuation fault represents an actual adverse condition. An indication/annunciation fault represents a "bad reading" or a "false alarm." Symptom recognition is basic data gathering. Identify and record the immediate and associated alarms or actuations, process conditions, systems testing, and any unusual sights, sounds, or smells at the time of the event. Resources for this data include first-out alarm recorders, charts and data logs, operator's logs, and on-shift operator interviews. Solicit and record as much information as possible. Once all the conditions are documented, systematically eliminate proven nonrelevant conditions. The level control system in Figure 13.1-1 is indicating a level low alarm. Identified symptoms include low level alarm in, high-level alarm not in, sight glass indicating low level, level control valve in the open position, level controller set-point is set at 50% and controller output gauge indicates 100% output to valve. Instrument air supply valves are open to the controller and valve. The solenoid is emitting a buzzing sound and venting air. The support documentation available is a system logic diagram, Figure 13.1-2, and instrument scale sheets, Figure 13.1-3.

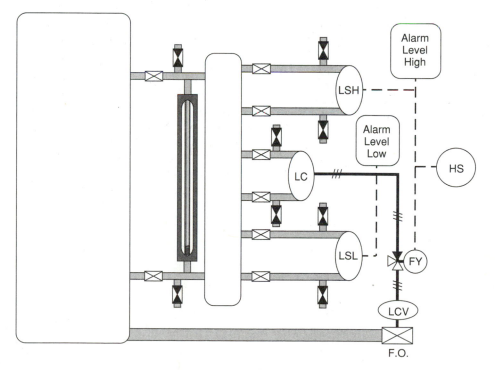

Figure 13.1-1 *Basic Level Control System*

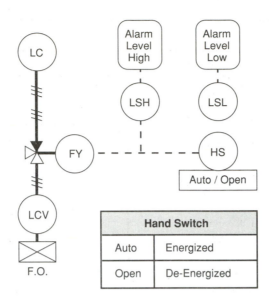

Figure 13.1-2 *Level System Logic Diagram*

LSL Scale Data Sheet	
Trip On Decrease @ 12" H_2O	
Reset	N/A
Contacts	N.C. & Com.
Function	Level Low Alarm

LSH Scale Data Sheet	
Trip On Increase @ 26" H_2O	
Reset	N/A
Contacts	N.C. & Com.
Function	Level High Alarm
	Fail-Open LCV

LC Scale Data Sheet	
Input	Output
-7" H_2O	3 psig
0" H_2O	9 psig
+7" H_2O	15 psig
Proportional	100% Direct
Set-Point	50%
Reset	Off
Rate	N/A

LCV Scale Data Sheet	
Input	Position
3 psig	0.0 Inches
9 psig	1.5 Inches
15 psig	3.0 Inches
Air Fail-Open	Volt Fail-Open
Benchset Closed	15 psig
Benchset Open	3 psig

Figure 13.1-3 *Level System Scale Sheets*

13.2 Symptom Analysis

Once the symptoms are identified, develop a list or chart and begin the analysis. The expectation of this step is to reduce an inclusive list of possible faults to an exclusive list of probable faults, such as Figure 13.2-1. Symptom analysis begins as an inductive process. For

Inductive		Deductive	
Symptom	**Possible Cause**	**Fault Reduction**	**Possible Faults**
Tank Level Low Alarm	Tank Level Low Level Switch Failure	Tank Level low Alarm Valid	
Sight Glass Level Low	Tank Level Low Level Valve Open	Tank Level Low Alarm Valid	
Valve Open	Hand Switch in Open Controller Output Low Instrument Air Off Valve Failure	Hand Switch in Auto Controller Output 100% Isolation Valves Open	Valve Failure
Controller	Set-Point Low Controller Failure Instrument Air Off	Set-Point 50% Controller Output 100%	
Hand Switch	Hand Switch in Open Hand Switch Failure	Hand Switch in Auto Solenoid Energized (Hot)	
Solenoid	Signal to Open Solenoid Failure	Solenoid Energized (Hot)	Solenoid Failure
Instrument Air	Isolated at Controller Isolated at Valve	Isolation Valves Open	

Figure 13.2-1 *Sympton Recognition and Analysis*

every single symptom, develop an expansive list of possible single point failures. Using a deductive process, review the list of single point failures and eliminate those that are disproved by reviewing the available information. Eliminate failures that are not common to all symptoms. The remaining single point failures that could affect the common components are probable faults. If all electrical components in a private residence fail simultaneously, it is possible that each individual device independently failed in unison. It is more probable that the main breaker tripped. If all kitchen appliances fail simultaneously, but the television remains operational, it is not possible for the main breaker to have tripped. It is probable that the breaker for the kitchen feed has tripped. In the working example, control valve fault and solenoid fault are identified as probable faults.

13.3 System Manipulation

The third phase is system manipulation. Proper system manipulation often includes recreating known conditions at the time of the event, or judiciously manipulating the system and monitoring component response to controlled changes. It is often helpful to install portable

Component	Manipulation	Results	Probable Faults
Tank Level Low Alarm Sight Glass	Manually Close Valve	Level Rises / Alarm Clears	Note
Valve	Hand Switch to OPEN Hand Switch to AUTO Drive Valve w/Controller Fail / Apply Air Jumper	No Change No Change No Change Valve Strokes w/Remote Air	Undetermined Undetermined Undetermined None
Controller	Lower Set-Point Match Set-Point to Input Raise Set-Point	Controller Output 100% Controller Output 50% Controller Output 0%	None
Hand Switch	Hand Switch in OPEN Hand Switch in AUTO	Solenoid De-Energized Solenoid Energized	None
Solenoid	Energize Solenoid Solenoid Failure	No Change	Solenoid Failure
Instrument Air	Check Air at Controller Check Air at Positioner Check Air at Solenoid Check Air from Solenoid	Air Present Air Present Air Present Air Not Present	None None None Solenoid Failure

Figure 13.3-1 *System Manipulation*

recorders or data loggers to capture real-time synchronized data. The expectation of system manipulation is to prove specific elements of a system good and reduce the list of multiple probable faults to a single point failure. Particular attention is directed to the conditions or components identified in the symptom analysis phase. Figure 13.3-1 indicates a solenoid failure is identified as a probable fault through system manipulation.

13.4 Fault Validation

The final phase of troubleshooting is fault validation. The expectation of fault validation is to distinguish and confirm the specific component or condition responsible for the problem identified. Symptom recognition, symptom analysis, and system manipulation lead to the isolation and identification of a specific suspected fault. Fault validation includes the measurement, monitoring, and analysis techniques required to positively identify and document the failure. Fault validation requires a working knowledge of the system and access to reliable prints and drawings. A system block diagram is developed that identifies All system inputs and all system outputs. Each "block" in the system is related to an AND gate. Systematically observe for and locate known good conditions. If the output of any block is good then all inputs to that block are considered good. If all the inputs to a block are good

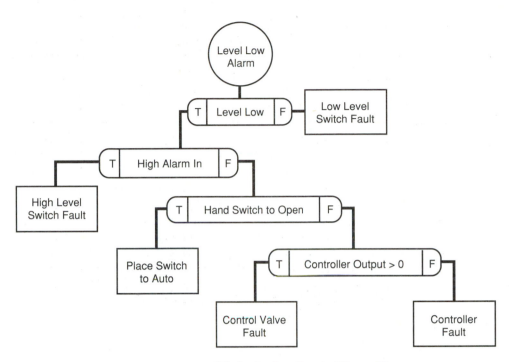

Figure 13.4-1 *High-Order Logic Flow Chart*

and any of the outputs are bad then that block is considered faulty. Half-split the system with each observation. Half-splitting is the process of successively dividing a system with each observation until the fault is bracketed. Visualize the signal path traveling through

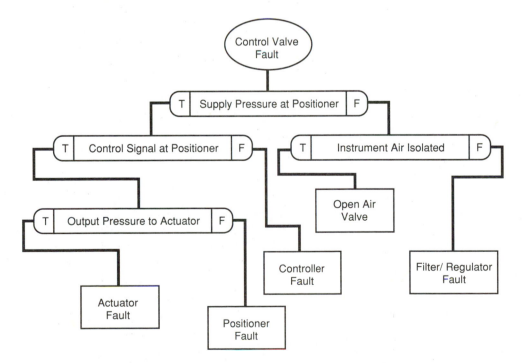

Figure 13.4-2 *Low-Order Logic Flow Chart*

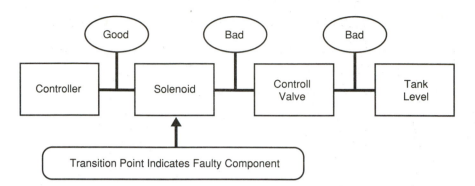

Figure 13.4-3 *Half-Split Bracketing*

several different modules from left to right. Locating a correct signal at the center of the flow path eliminates all upstream modules, those to the left, as a failure source. The next measurement is taken midway between the remaining suspect modules, those on the right. If the correct signal is not observed, the faulty component is bracketed between the two measurements. The third measurement is taken midway between the first and second. Continue locating known good conditions until the faulty component is identified. The transition point of a known good condition to an erroneous condition indicates a faulty

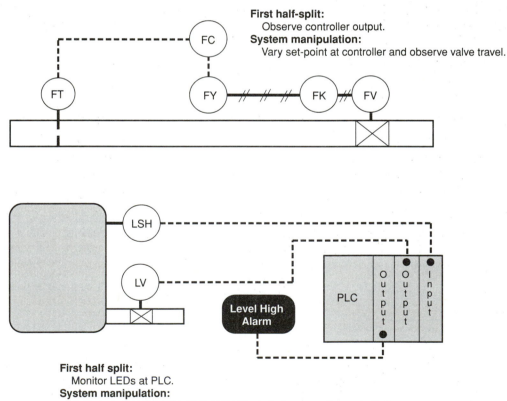

Figure 13.4-4 *Typical Half-Split Techniques*

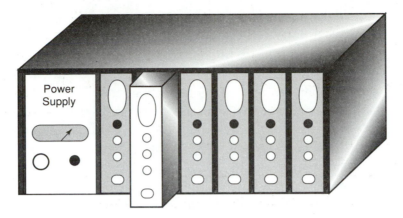

Figure 13.4-5 *Power Supply/Ground Fault Isolation*

component. High order logic flow charts, Figure 13.4-1, isolate the failure to a control valve fault. Low-order logic flow chart, Figure 13.4-2, isolates the failure to a controller fault. System manipulation proved the controller good. High-order logic indicated control valve malfunction and low-order logic eliminated all other valve components. The only component between the controller and valve is the solenoid and system manipulation verified the solenoid control signal changing state at the solenoid. A block diagram is developed of the questioned components and half-splitting isolates the solenoid as the single point failure, Figure 13.4-3. Figure 13.4-4 demonstrates common half-split techniques.

Certain conditions such as power supply failures and ground faults present a unique challenge. If the entire system is de-energized, the traditional approaches of symptom recognition and system manipulation are impractical. The most direct approach is to unload the bus, restore power, and then systematically restore segments of the system until the ground or excess load is detected. Clamp-on ammeters are quite useful to locate high current drains. Disconnect and check the cable runs with a megohmmeter. Check each conductor to each other conductor and to ground. Verify new installations for continuity. This is accomplished with a multimeter in the ohms position or a telephone headset designed for such purposes. Figure 13.4-5 shows a card partially removed from the cabinet's backplane. The faulty component is then replaced and analyzed to validate the root cause of the failure.

13.5 Change Analysis

Change analysis does not readily identify any particular fault, but in many cases serves to bracket preliminary investigations to a defined arena and save time. The perspective of review is to determine the last point in time that the system performed flawlessly and then develop a list of activities that have occurred since that time. The activities listed are directly related to the system and system components, physically near, or electrically congruent to the system investigated. If a machine that performed flawlessly for an extended time is taken out of service for a routine preventive maintenance, and exhibits problems when placed back into service, the details of the preventive maintenance activity are suspect. The validity and usefulness of change analysis is contingent to the number of changes made. The optimal situation is to integrate each change in discrete steps, and validate proper equipment performance after each change. If it is impractical to effect individual aspects of change, attempt to divide the system into separate functional subunits and confine the scope of changes to individual subunits. This writer once encountered a temperature-monitoring loop that occasionally dropped twenty degrees below the actual process temperature. The point would remain low anywhere from five minutes to twelve hours and then return to the correct reading. After investigating and verifying all associated hardware in the loop, no problem was found. Change analysis revealed a single activity in the immediate area; an overhead steam line had been reinsulated. Further research linked the temperature change to incidents of rainfall greater than 1/2" per hour. Fortunately, an opportunity to observe the combined conditions shortly presented itself. As unlikely as it seemed, the insulation was installed in such a manner to channel excess rainwater directly onto the thermocouple thermo well. The steam line insulation was modified and the temperature indication error never recurred. This is an example of compounded conditions. Neither circumstance could independently create an adverse condition, but when multiple conditions occurred simultaneously, an error condition existed.

■ Chapter Thirteen Summary

- Troubleshooting is the ordered process of investigating, identifying, and eliminating a fault or combination of faults responsible for undesirable system conditions.
- Troubleshooting restores a system to desirable operating conditions and eliminates probable replicate failures.
- Eliminating or masking symptoms without correcting the root cause creates a potentially dangerous and expensive situation.
- Failures are either random or systematic. Random failures are spontaneous and generally hardware related. Systematic failures are caused by underlying faults of design or application.
- Faults are either permanent faults or random faults. A permanent fault creates adverse conditions that persist until the fault is remedied. Random faults create temporary adverse conditions and are difficult to locate. Random faults are often linked to other outside influences such as radio frequency transmission, cross talk, temperature, and moisture.
- Symptom recognition is basic data gathering.
- The primary expectation of symptom recognition is to determine the precise validity and identity of the undesirable condition.
- Resources for symptom data include first-out alarm recorders, charts and data logs, operator's logs, and on-shift operator interviews.
- Symptom analysis develops a list of possible single point failures for every single symptom and then using a deductive process, reviews the list and eliminates those

that are disproved. The expectation of this step is to reduce an inclusive list of possible faults to an exclusive list of probable faults.

- A possible fault is a single point failure that is capable of generating some of the observed symptoms.
- A probable fault is a single point failure that is capable of generating all observed symptoms.
- System manipulation is the judicious operation and monitoring of component response to controlled changes.
- The expectation of system manipulation is to prove specific elements of a system good and reduce the list of multiple probable faults to a single point failure.
- Alarms, portable recorders, and data loggers capture real-time synchronized system manipulation data.
- Fault validation is the measurement, monitoring, and analysis techniques to positively identify and document a failure.
- The expectation of fault validation is to distinguish and confirm the specific component or condition responsible for the problem identified.
- Fault validation requires a working knowledge of the system and access to reliable prints and drawings.
- Half-splitting is the process of successively dividing a system with each observation until the fault is bracketed.
- The most direct approach to troubleshoot power supply and grounding faults is to unload the bus, restore power, and then systematically restore segments of the system until the ground or excess load is detected.

Chapter Thirteen Review Questions:

1. Define troubleshooting.

2. State the expectation of troubleshooting.

3. List the adverse effects of improper troubleshooting.

4. List the four phases of troubleshooting.

5. Describe the symptom recognition phase of troubleshooting.

6. State the expectation of the symptom recognition phase of troubleshooting.

7. List several resources for symptom recognition data.

8. State the expectation of the symptom analysis phase of troubleshooting.

9. Describe the difference between a possible fault and a probable fault.

10. Describe the system manipulation phase of troubleshooting.

11. State the expectation of the system manipulation phase of troubleshooting.

12. List several resources for system manipulation data.

13. Describe the fault validation phase of troubleshooting.

14. State the expectation of the fault validation phase of troubleshooting.

15. List several resources for fault validation data.

16. Describe the half-split system of fault validation.

17. Describe the technique to isolate power supply and grounding faults.

Chapter 14

Safety, Certainty, and Professionalism

OBJECTIVES

Upon completion of Chapter Fourteen the student will be able to:

- Interpret an MSDS and determine appropriate safety precautions.
- Interpret a DOT placard and determine the type and level of hazard.
- State the minimum percent oxygen required to support human life.
- State the purpose of explosionproof electrical devices.
- State the hazards of working in a confined space.
- List potential work area hazards and remediation techniques.
- List items used to reduce or eliminate environmental risk.
- List challenges introduced by safety equipment.
- Identify the individual or organization ultimately responsible for your safety.
- State the purpose of, and areas covered by, a team briefing.
- List actions required prior to initiating an activity.
- List techniques used to enhance clear communication.
- Differentiate between the terms may, should and shall.
- Differentiate between the terms note, caution, and warning.
- State the phonetic alphabet, and describe three-part communication.
- Define abatement, anomaly, and error precursor.
- State the impact of disregarding anomalies and error precursors.
- Describe the proper lockout/tagout procedure according to OSHA.
- Define process restoration,.
- State the most significant safety contribution instrument technicians can make.
- State the benefits of the application of professional attitudes and actions.

14.0 Introduction

Safety awareness is undoubtedly the most significant aspect of professionalism. A momentary lapse in safety awareness can lead to a lifetime of regret. Processing facilities are alive with activity. Technicians routinely work with, on, and around high-energy systems. Process media may often contain hazardous materials. Conditions are constantly changing. Previously clear walkways are obstructed in a matter of minutes. Risk potentials change unexpectedly. Each work group is naturally focused on their primary task and can momentarily lose awareness of their surroundings. This can create a dangerous situation. Certain activities are recognized as having high potential to create this condition and as such are routinely performed only with a dedicated observer. These activities include welding, entering confined spaces, and working in high-radiation areas. It is not always practical to have a designated observer assigned to every ongoing task. Much of the time technicians must accept the responsibility to simultaneously perform work and maintain safety awareness.

14.1 Hazardous Material Safety

A high-energy process is any process that contains significant amounts of latent destructive or dangerous force. High-pressure fluids and gasses, high-temperature products, oxidizers, flammable and explosive liquids and gasses are examples of a high-energy process. High-pressure fluids and gasses present an immediate physical danger in an uncontrolled release. High-temperature products can cause serious and even fatal burns. Oxidizers react violently to air and other chemicals and can generate explosive pressures almost instantly. Combustible gasses and vapors have an upper explosive limits (UEL) and lower explosive limits (LEL). The concentration levels between the upper and lower limit are the concentrations that will support flammable or explosive reactions. Work activity near explosive or flammable gas and vapors requires the use of arc-tight explosionproof lighting. The housing and connections are constructed to prevent any gas, vapor, mist, or fume from reaching the electrical contacts inside. Special nonsparking tools are also available. Inspect the area for combustible or flammable materials. Locate and inspect fire extinguishers near the work area. Remediation could include draining, venting, and flushing high-energy or hazardous piping. Employers maintain a safety tag or lockout process. Always respect and comply with the process. If the tagout is not adequate, have it corrected. Tampering with tagouts and lockout devices can cause serious injury and death.

Hazardous material should be clearly labeled with a fire diamond, Figure 14.1-1. The fire diamond identifies the type and severity of hazard presented by the labeled material. Hazardous processes include acids, caustics, and carcinogens.

Federal regulation OSHA 29 CFR 1910 contains the United States Government's Code of Federal Regulations concerning the care, use, shipping, labeling, and disposal requirements of hazardous materials. Obtain and review the Material Safety Data Sheet (MSDS) for all

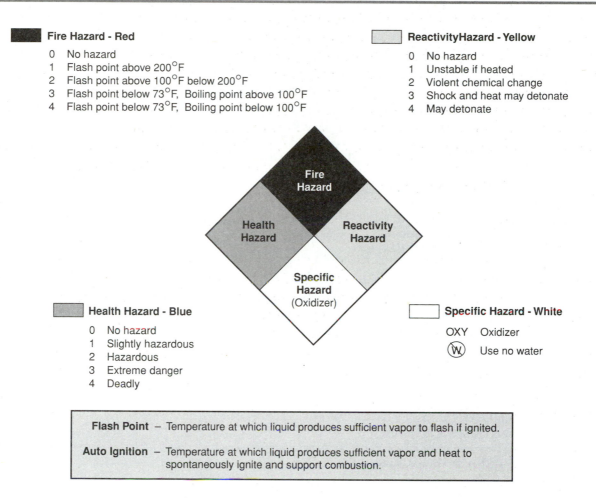

Fire Hazard - Red

0 No hazard
1 Flash point above 200°F
2 Flash point above 100°F below 200°F
3 Flash point below 73°F, Boiling point above 100°F
4 Flash point below 73°F, Boiling point below 100°F

Reactivity Hazard - Yellow

0 No hazard
1 Unstable if heated
2 Violent chemical change
3 Shock and heat may detonate
4 May detonate

Health Hazard - Blue

0 No hazard
1 Slightly hazardous
2 Hazardous
3 Extreme danger
4 Deadly

Specific Hazard - White

OXY Oxidizer
Ⓦ Use no water

Flash Point – Temperature at which liquid produces sufficient vapor to flash if ignited.

Auto Ignition – Temperature at which liquid produces sufficient vapor and heat to spontaneously ignite and support combustion.

Figure 14.1-1 *Fire Diamond*

potential products involved with the work activity. Many common cleaning solvents and lubricants also require special use, labeling, handling, and disposal considerations. The MSDS provides specific exposure limits, safety equipment, and first aid requirements for each product (see Figure 14.1-2).

Occupational exposure standards are established by the Occupational Safety and Health Administration (OSHA) as permissible exposure limits (PEL). Personal exposure limits are defined for four distinct levels of permissible exposure times: eight-hour time-weighted average concentration (TWA), fifteen-minute time-weighted average concentration, or short-term exposure (STEL), ceiling concentration (C), and immediately dangerous to life and health (IDLH). TWA exposure is the maximum concentration exposure allowed for a worker to receive in a period not to exceed eight hours in each twenty-four period. STEL concentration assumes an exposure not exceeding fifteen minutes within a twenty-four hour period. Ceiling concentration is the amount of hazardous material at which no exposure is permitted. IDLH is a concentration level that produces death or permanent adverse health affects with not more than a thirty-minute exposure time.

===

Section 1. CHEMICAL PRODUCT SECTION
Product Name: Isopropyl Alcohol
Product Number: XX-XXX
General Use: Cleaner/Degreaser
MANUFACTURER: XXX For Chemical Emergency, Spill, Leak, Fire
Exposure, or Accident Call DAY OR NIGHT 1-800-XXX-XXXX.

===

Section 2. COMPOSITION / INFORMATION ON INGREDIENTS
CHEMICAL C.A.S Isopropanol 67-63-0.   Number Weight % 100
OSHA HAZARDOUS COMPONENTS (29 CFR 1910.1200):
Exposure Limits 8 Hours TWA (PPM)
OSHA PEL ACGIH TLV 400 PPM/ 500STEL

===

Section 3. HAZARD IDENTIFICATION
Emergency Overview:
Potential Health Effects:
INHALATION: Major route of exposure. Vapor is heavier than air and can cause suffocation by reducing the available oxygen for breathing. Breathing high concentrations of vapor may cause light-headedness, giddiness, shortness of breath, confusion, and may lead to narcosis, cardiac irregularities, unconsciousness or even death.
EYES: Liquid can cause slight, temporary irritation with slight temporary corneal injury. Vapors can irritate eyes.
SKIN: Product may cause skin irritation. Repeated of prolonged contact may cause redness with burning, drying, and cracking of skin. This product may be absorbed through the skin.
INGESTION: This product is toxic by ingestion. Ingestion may cause irritation of the digestive tract. Signs of nervous system depression (e.g. drowsiness, dizziness, loss of coordination, and fatigue). Nausea and vomiting will most likely occur. This product has not been classified as a carcinogen or a probable carcinogen by NTP, IARC, or OSHA.

===

Section 4. FIRST AID MEASURES
Inhalation:
Move to fresh air in case of accidental inhalation of vapors. If victim has stopped breathing, give artificial respiration. Call for prompt medical attention.
Eye Contact:
Flush eyes with large amounts of water for 15 minutes or until irritation subsides. If irritation persists, get medical attention.
Skin Contact:
Remove contaminated clothing (including shoes) and wash before reuse. Flush with large amounts of water. Use soap if available. If irritation persists, seek medical attention.
Ingestion:
Do not induce vomiting unless directed by a physician. If conscious and alert, give two glasses of water. Seek medical attention immediately.

===

Section 5. FIRE FIGHTING MEASURES
Flash Point & Method: 53 F, TCC Method
Flammable Limits: LEL: 2.0   UEL: 12.0
FIRE FIGHTING INSTRUCTIONS:
Fire fighters should wear self contained, positive-pressure breathing apparatus and avoid skin contact.
FIRE FIGHTING EQUIPMENT:
Water, foam, dry chemical, carbon dioxide.
HAZARDOUS COMBUSTION PRODUCTS:
Smoke, fumes and oxides of carbon.

Figure 14.1-2 *Material Safety Data Sheet (Sample Only – Not For Use)*

Adverse health effects are acute and/or chronic. Acute effects result from a short-term exposure and develop symptoms rapidly. Chronic effects result from sustained long-term exposure to low levels and develop symptoms slowly over time.

Prior to entering an area or engaging in an activity evaluate the proposed work and inspect the surroundings. Note all obvious hazard potentials. These include high-energy/hazardous processes, high-voltage equipment, rotating equipment, slip/trip hazards, overhead work, and other work groups in the area, vehicular traffic, and environmental conditions. Resolve these conditions before proceeding. Locate and ensure operability of safety eyewash and shower stations near the work area. Locate and test communication equipment near the work area. Notify others in the area of activities, interrupt overhead work, insure scaffolds and walks have adequate aprons or skirting in place, and ensure ladders are used properly and tied off. Attach lanyards on tools or secure any items that could become a hazardous falling object. Remove or mark slip/trip hazards, and maintain good housekeeping and personal hygiene practices at all times. Four major paths of exposure include inhalation, ingestion, direct contact with skin, and exposure to the eyes. While personal protective equipment can protect the worker from many hazards, it is critical that the worker employ proper hygiene at all times. When removing safety equipment ensure the articles are removed in such a manner as to not cross-contaminate the unprotected areas of the body. Thoroughly cleanse the hands before eating, using tobacco, or touching the face or other areas of the body.

Respiratory protection often requires a respirator. The three basic types of respirators are self-contained breathing apparatus (SCBA), supplied air respirators (SAR), and air-purifying (filter or cartridge) respirators. SCBA and SAR devices provide independent breathing air and are approved for oxygen-deficient environments. Positive-pressure SCBAs are suitable for oxygen-deficient environments and atmospheres immediately dangerous to life and health. Air-purifying respirators do not supply independent breathing air and are not suitable for oxygen-deficient or IDLH atmospheres. Full-face filter respirators offer protection against hazardous gas, mist, fumes, and vapors, but do not compensate for oxygen-deficient environments. Respirators have a protection-factor number. This number represents the effectiveness of the respirator. Cartridges and filters are designated for specific hazards. The effectiveness of canisters and cartridges deteriorates over time, especially in high humidity. Prior to donning the respirator, inspect it for damage, particularly around the sealing area. Perform a negative pressure test by blocking the inlet with your hand and attempting to inhale. Do not use the respirator unless it holds a negative pressure.

Confined and enclosed spaces present a very real hazard. These atmospheres are often oxygen deficient less than 20.9%. Employers maintain a specific confined entry permit process. Do not take chances; adhere to the permit guidelines at all times. Open the space and ventilate sufficiently before entering. Constant ventilation is required as substances continue to leach from the walls and corrosive activity continues to deplete the oxygen. Never enter an enclosed or confined space without an oxygen monitor. Do not attempt to enter and remove anyone from a confined space without first notifying others, attaching a recovery line, and obtaining supplied breathing air. For every person that has died in an enclosed space, many more were lost during futile rescue attempts.

Many adverse environments are eliminated or minimized by engineering controls. Ventilation, lighting, sound reduction materials, guardrails, and drainage all reduce hazard levels. Unavoidable adverse environments require specialized equipment. Safety harnesses, face shields, SCBA, cooling vest, gas or oxygen monitors, ear protection, and other equipment can reduce or eliminate the obvious environmental risk, but introduce other considerations. Peripheral vision, hearing, mobility, and stamina are often compromised by the use of safety equipment. Evaluate the extent of the stresses involved and develop acceptable stay and recovery times. Proper training, equipment, and culture can minimize the frequency and extent of adverse conditions, but the ultimate responsibility for safe work activity resides in the individual.

14.2 Certainty

Safe activities are controlled activities. It is critical to maintain control of evolutions at all times. Certainty is a critical component of safety. Prior to any activity, develop a plan of action. The plan of action is communicated to all participants in a team briefing. The purpose of the team briefing is to develop and communicate a clear overview of the entire evolution. The briefing should clearly designate specific individuals to perform each task. Key roles identified include command and control, communications, and support. The brief precisely identifies what equipment is manipulated, what process is affected, and the expected results. The specific expectations of the evolution including the purpose, impact, and relevance are also stated. Critical steps and checkpoints are designated. Potential negative results are identified and a plan of action to prevent or mitigate potential process transients is discussed. The location of personnel and components involved is established, and a synchronized time line for the initiating event, each critical step, and projected completion is fixed. Prior to initiation of the evolution, perform a physical walk down of the area and equipment. Locate and verify the operability of safety support devices such as fire extinguishers, safety showers, and eyewash stations. Isolate, vent, and drain the system when practical. Notify all at-risk personnel in the area or erect exclusion barriers.

Use precise communication techniques during the evolution. Avoid excessive use of acronyms and slang. Carefully select and employ discrete terminology; ambiguous words and phrases are easily misunderstood. The term *may* delegates permission but does not imply a recommendation. The term *should* implies a recommendation but not a specific requirement. The term *shall* conveys an explicit requirement. *Verify* means to observe and record only, while the term *ensure* implies an action if required. The term *note* is a recommendation to take notice, observe, or direct attention to something. The term *caution* is a statement of concern and suggests vigilance. The term *warning* implies that danger is imminent. The phonetic alphabet, Figure 14.2-1, and three-part communications are often

International Radio Alphabet			
Letter	Word	Letter	Word
A	Alpha	N	November
B	Bravo	O	Oscar
C	Charlie	P	Papa
D	Delta	Q	Quebec
E	Echo	R	Romeo
F	Foxtrot	S	Sierra
G	Golf	T	Tango
H	Hotel	U	Uniform
I	India	V	Victor
J	Juliet	W	Whiskey
K	Kilo	X	X-ray
L	Lima	Y	Yankee
M	Mike	Z	Zulu

Figure 14.2-1 *International Radio Alphabet*

used to ensure an accurate exchange of information. Three-part communication consists of the sender stating the original phrase, the receiver restating the original phrase, and the sender acknowledging the restatement as correct. For example, the controller states "Take switch whiskey zulu tango to the close position." The support technician replies, "I understand take switch whiskey zulu tango to the close position." The controller acknowledges with "that is correct". Once the switch is manipulated, the support technician states, "Switch whiskey zulu tango is in the close position." The controller repeats back "I understand, switch whiskey zulu tango is in the close position." The support technician then acknowledges, "That is correct."

14.3 Professionalism

Process control systems are originally constructed from a set of design drawings. The design drawings are used to establish the set-points and operational characteristics of all associated instrumentation. It is common to encounter limited discrepancies during the start-up of most systems. All deviations and required modifications must be documented and transferred as revisions to the original design drawings. Unfortunately, the revision process occasionally fails, leaving the technician with incomplete, or worse incorrect, information. A professional instrument technician must always maintain a questioning attitude. Avoid complacency at all times. Some procedures are so thoroughly documented and commonly performed that all of the "bugs" are ironed out. A technician may be lured into a comfortable state of cookbook mentality. Never allow yourself to become a mindless automaton. A state of increased awareness is required throughout the entire evolution. All participants must maintain a questioning attitude. An operation technician that has been cycling pumps on the second level for most of the shift may habitually travel back to the second level when requested to cycle a pump located on the third level. Perform a self-check before actuating equipment or manipulating controls. To perform a self-check, stop and consciously verify your location and the identification of the component. Extremely critical manipulations often require a peer check as well. Communicate identified error-likely situations and error precursors immediately. Evaluate all anomalies observed and ensure successful completion of each step before continuing. Error precursors are often called near misses or close calls. Failure to determine and correct the root cause of a near miss permits the near miss event to persistently recur until calamity eventually results.

14.4 Lockout/Tagout

Instrumentation technicians routinely perform maintenance on high-energy and hazardous systems. An energy source is any source of electrical, mechanical, hydraulic, pneumatic, chemical, thermal, or other energy. Their work activity has the potential to start and stop equipment, open and close valves, and disable critical alarm and trip functions. OSHA defines the lockout/tagout procedure requirements necessary to perform this work in a safe manner in the controlled documents "The Control of Hazardous Energy (Lockout/Tagout), Title 29 CFR 1910.147, 269, and 333." An energy control program consists of energy control procedures, employee training, and periodic inspections to ensure that before any employee performs servicing or maintenance on equipment where the unexpected energizing, start-up or release of stored energy could occur and cause injury, the machine or equipment shall be isolated from the energy source and rendered inoperative. Energy control procedures are utilized for the control of potentially hazardous energy when employees are engaged in covered activities. A covered activity is any activity that fails to meet all of the following eight exceptions:

(1) The equipment has no potential for stored or residual energy or reaccumulation of stored energy after shutdown which could endanger employees.

(2) The equipment has a single energy source that can be readily identified and isolated.

(3) The isolation and locking out of that energy source will completely de-energize and deactivate the machine or equipment.

(4) The machine or equipment is isolated from that energy source and locked out during servicing or maintenance.

(5) A single lockout device will achieve a locked-out condition.

(6) The lockout device is under the exclusive control of the authorized employee performing the servicing or maintenance.

(7) The servicing or maintenance does not create hazards for other employees.

(8) The employer, in utilizing this exception, has had no accidents involving the unexpected activation or reenergization of the machine or equipment during servicing or maintenance.

The energy control procedure outlines the rules and techniques utilized for the control of hazardous energy. It includes specific procedural steps for shutting down, isolating, blocking, and securing equipment to control hazardous energy. The placement, removal, and transfer of lockout devices or tagout devices and the responsibility for them are also covered. Specific requirements are stated for testing equipment to verify the effectiveness of lockout devices, tagout devices, and other energy control measures. It also defines protective materials and hardware such as locks, tags, chains, wedges, key blocks, adapter pins, self-locking fasteners, or other hardware that shall be provided by the employer for isolating, securing or blocking of machines or equipment from energy sources.

Acceptable energy-isolating devices are mechanical devices that physically prevent the transmission or release of energy such as a manually-operated electrical circuit breaker, a disconnect switch, a line valve, a block, and any similar device used to block or isolate energy. Pushbuttons, selector switches, and other control circuit-type devices are not energy-isolating devices.

A lockout device is a device that uses a positive means such as a lock, either key or combination type, to hold an energy-isolating device in the safe position and prevent the energizing of a machine or equipment. The placement of a lockout device on an energy-isolating device, in accordance with an established procedure, ensures that the energy-isolating device and the equipment being controlled cannot be operated until the lockout device is removed. A tagout device is a prominent warning device, such as a tag and a means of attachment, which can be securely fastened to an energy-isolating device in accordance with an established procedure, to indicate that the energy-isolating device and the equipment being controlled may not be operated until the tagout device is removed. Lockout and tagout devices shall be capable of withstanding the environment to which they are exposed, constructed, and printed so that exposure to weather conditions or wet and damp locations will not cause the tag to deteriorate or the message on the tag to become illegible, and shall not deteriorate when used in corrosive environments. They shall also be standardized in color, shape, or size, print, and format. Lockout devices shall be substantial enough to prevent removal without the use of excessive force or unusual techniques, such as with the use of bolt cutters or other metal cutting tools. Tagout devices shall be substantial enough to prevent inadvertent or accidental removal. Tags and their means of attachment must be made of materials that will withstand the environmental conditions encountered in the workplace. Tags must be securely attached to energy-isolating devices so that they cannot be inadvertently or accidentally detached during use. Lockout devices and tagout devices shall indicate the identity of the employee applying the device(s). Tagout devices shall warn against hazardous conditions if the machine or equipment is energized and shall include a legend such as the following:

Do Not Start. Do Not Open. Do Not Close. Do Not Energize. Do Not Operate.

Tags are essentially warning devices affixed to energy-isolating devices and do not provide the physical restraint on those devices that is provided by a lock. Tags may evoke a false sense of security and their meaning needs to be understood as part of the overall energy control program.

When a tag is attached to an energy-isolating means it is not to be removed without authorization of the person responsible for it and it is never to be bypassed, ignored, or otherwise defeated. In order to be effective, tags must be legible and understandable by all employees whose work operations are or may be in the area.

Lockout/tagout procedures delineate a safe and orderly method for securing a process or equipment. Before an employee turns off a machine or equipment, the employee shall have knowledge of the type and magnitude of the energy, the hazards of the energy to be controlled, and the method or means to control the energy. The machine or equipment shall be turned off or shut down using the procedures established for the machine or equipment. An orderly shutdown must be utilized to avoid any additional or increased hazard(s) to employees as a result of the equipment stoppage. All energy-isolating devices that are needed to control the energy to the machine or equipment shall be physically located and operated in such a manner as to isolate the machine or equipment from the energy source(s). Authorized employees shall affix lockout or tagout devices to each energy-isolating device. Lockout devices shall be affixed in a manner that will hold the energy-isolating devices in a "safe" or "off" position. Tagout devices shall be affixed in a manner that clearly indicates the operation or movement of energy-isolating devices from the safe or off position is prohibited. Where tagout devices are used instead of locks, the tag attachment shall be fastened at the same point at which the lock would have been attached. Where a tag cannot be affixed directly to the energy-isolating device, the tag shall be located as close as safely possible to the device, in a position that will be immediately obvious to anyone attempting to operate the device. Following the application of lockout or tagout devices, all potentially hazardous stored or residual energy shall be relieved, disconnected, restrained, and otherwise rendered safe. If there is a possibility of reaccumulation of stored energy to a hazardous level, isolation shall be continued until the servicing or maintenance is completed, or until the possibility of such accumulation no longer exists.

◆ CAUTION ◆ Prior to starting work on machines or equipment that have been locked out or tagged out, verify that adequate isolation and deenergization of the machine or equipment have been accomplished.

Before lockout or tagout devices are removed and energy is restored to the machine or equipment, inspect the work area and ensure that nonessential items have been removed and that machine or equipment components are operationally intact. Ensure that all employees have been safely positioned or removed from the work area. After lockout or tagout devices have been removed and before a machine or equipment is started, notify employees that the lockout or tagout device(s) have been removed. The employee who applied the device shall remove each lockout or tagout device from the energy-isolating device. In the event the authorized employee who applied the lockout or tagout device is not available to remove it, that device may be removed under the direction of the employer, provided that: the authorized employee who applied the device is not at the facility; all

Lift/Land Documentation Table				
Device	**Device Description**	**Termination Point**	**Lifted**	**Landed**
Lead	Cable 123XYZ (WH)	Cabinet ZLC 123 TB - J Terminal 12	initials mm/dd/yy	initials mm/dd/yy
Lead	Cable 123XYZ (BK)	Cabinet ZLC 123 TB - J Terminal 13	initials mm/dd/yy	
Shield	Cable 123XYZ (SH)	Cabinet ZLC 123 TB - J Terminal 14	initials mm/dd/yy	initials mm/dd/yy
n/a	n/a	n/a	n/a	n/a
Comments:				

A quick review of the lift / land log reveals that lead, cable 123XYZ (BK), was lifted, but has not been re-landed.

Figure 14.4-1 *Lift/Land Documentation Sheet*

reasonable efforts are made to contact the authorized employee to inform him/her that his/her lockout or tagout device has been removed, and that the authorized employee has this knowledge before he/she resumes work at that facility.

While equipment motor controls, breakers, fuses, and manual valve positions are often controlled under a safe tagout program, the technician may manipulate equipment, set-points, and wiring as part of the work process. Prior to performing these activities ensure that the lockout/tagout program does not cover them. Control device and wiring configurations with a lift-and-log, and document the restoration process. See Figure 14.4-1. Verify all components and conditions are returned to their desired configuration and the system left in a safe condition. While no profession is absolutely devoid of all risk, the most significant safety contribution instrument technicians can make is to accept ownership and personal accountability for their attitudes and actions. The judicious and dedicated application of professional attitudes and actions creates a safe and productive working environment for everyone.

■ Chapter Fourteen Summary

- Prior to entering an area or engaging in an activity perform an inspection and evaluation of the surroundings. Note all obvious hazard potentials. Resolve these conditions before proceeding.

- Safety harnesses, face shields, SCBA, cooling vest, gas or oxygen monitors, and other equipment can reduce or eliminate environmental risk.

- Peripheral vision, hearing, mobility, and stamina or often compromised by the use of safety equipment.

- Proper training, equipment, and culture can minimize the frequency and extent of adverse conditions.

- The ultimate responsibility for safe work activity resides in the individual.

- The plan of action is communicated to all participants in a team briefing. The purpose of the team briefing is to develop and communicate a clear overview of the entire evolution.

- The brief precisely identifies equipment manipulated, process affected, and the specific expectations of the evolution. Critical steps and checkpoints are designated, and a plan of action to prevent or mitigate potential process transients is discussed. The location of personnel and components involved is established and a synchronized time line is fixed.

- Prior to initiation of the evolution, perform a physical walk down of the area and equipment. Isolate, vent, and drain the system when practical. Notify all at-risk personnel in the area or erect exclusion barriers.

- Avoid excessive use of acronyms and slang. The phonetic alphabet and three-part communications are often used to ensure an accurate exchange of information.

- Three-part communication consists of the sender stating the original phrase, the receiver restating the original phrase, and the sender acknowledging the restatement as correct.

- Error precursors are often called near misses or close calls.

- Failure to determine and correct the root cause of a near miss permits the near-miss event to persistently occur until calamity eventually results.

- Mitigation is an action that reduces or alleviates the adverse impact of an undesirable event.

- Abatement is an action that stops or eliminates an undesirable condition.

- An anomaly is an event that is irregular and deviant from normal conditions.

- The official guidelines for lockout/tagout requirements are located in OSHA document 29 CFR 1926.147, 269, and 333.

- There are eight parameters or conditions that determine if a tagout/lockout is required.

- Document the restoration process. Verify all components and conditions are returned to their desired configuration and the system is left in a safe condition.

- The most significant safety contribution operation technicians can make is to accept ownership and personal accountability for their attitudes and actions.

Chapter Fourteen Review Questions:

1. List potential work area hazards.

2. List remediation techniques used to reduce area hazards.

3. List items used to reduce or eliminate environmental risk.

4. List challenges introduced by safety equipment.

5. List approaches used to minimize adverse conditions.

6. Identify the individual or organization ultimately responsible for your safety.

7. State the method used to communicate a plan of action to all participants in a team.

8. State the purpose of a team briefing.

9. List areas covered in a team briefing.

10. List actions required prior to initiating an activity.

11. List techniques used to enhance clear communication.

12. State the phonetic alphabet.

13. Describe three-part communication.

14. Define mitigation.

15. Define abatement.

16. Define anomaly.

17. Define error precursor.

18. State the impact of disregarding anomalies and error precursors.

19. Explain the tagout/lockout requirements per OSHA.

20. State the most significant safety contribution that instrument technicians can make.

21. State the benefits of the judicious and dedicated application of professional attitudes and actions.

Index